System Identification

System Identification

Least-Squares Methods

T. C. Hsia
University of California, Davis

Lexington Books
D. C. Heath and Company
Lexington, Massachusetts
Toronto

Library of Congress Cataloging in Publication Data

Hsia, Tien C
   System identification.
   Includes index.
   1. System analysis. 2. Least squares. I. Title.
QA402.H76   003   75–3515
ISBN 0–669–99630–0

Copyright © 1977 by D.C. Heath and Company

All rights reserved. No part of this publication may be reproduced or transmitted in any form or by any means, electronic or mechanical, including photocopy, recording, or any information storage or retrieval system, without permission in writing from the publisher.

Published simultaneously in Canada

Printed in the United States of America

International Standard Book Number: 0–669–99630–0

Library of Congress Catalog Card Number: 75–3515

To Lillian, Lily, and Margaret

Contents

	Preface	xvii
Chapter 1	**Introduction**	1
	1.1 The Problem of System Identification	1
	1.2 Formulation and Classification of the System Identification Problem	2
	1.3 Parameter Estimation Methodology	4
	1.4 Organization of the Text	5
Chapter 2	**Representation of Dynamic Systems**	7
	2.1 Introduction	7
	2.2 Linear Difference Equation	7
	2.3 Weighting Sequence and Convolution	9
	2.4 State Variable Equation	11
	2.5 Concluding Remarks	15
Chapter 3	**Least-squares Theory**	17
	3.1 Introduction	17
	3.2 Least-squares Theory	17
	3.3 Statistical Properties of Least-squares Estimators	20
	3.4 Sequential Least-squares Estimation	22
	3.5 Multi-dependent-variable System	25
	3.6 Recursive Estimation for Increasing Parameter Numbers	27
	3.7 Real-time Least-squares Algorithm	29
	3.8 Concluding Remarks	31
Chapter 4	**Weighting Function Identification**	37
	4.1 Introduction	37
	4.2 The Identification Problem	38
	4.3 Least-squares Estimation	40
	4.4 Relationship to Cross-correlation Identification	43
	4.5 Optimum Input Signal	47
	4.6 Pseudo-random Binary Sequence	49

	4.7 On-line Least-squares Identification	55
	4.8 Multivariable System Identification	58
	4.9 Concluding Remarks	61
Chapter 5	**Linear Parametric Model Identification by Least Squares**	**67**
	5.1 Introduction	67
	5.2 The Basic Identification Problem	68
	5.3 Least-squares Solution	69
	5.4 Statistical Properties of Parameter Estimates	72
	5.5 On-line Least-squares Identification	75
	5.6 System Order Determination	76
	5.7 Real-time Identification	81
	5.8 Continuous Systems Identification	84
	5.9 Concluding Remarks	89
Chapter 6	**System Identification by Generalized Least Squares**	**97**
	6.1 Introduction	97
	6.2 Formulation of Noisy System Model	97
	6.3 Bias Problems Associated with Correlated Residuals	98
	6.4 Formulation of the Generalized Least-squares Problem	100
	6.5 Generalized Least-squares Estimation algorithm	101
	6.6 Remarks	106
	6.7 An Alternative GLS Solution Technique	111
	6.8 Instrumental Variable Method	119
	6.9 Concluding Remarks	122
Chapter 7	**Multistage Least-squares Identification Technique**	**125**
	7.1 Introduction	125
	7.2 MSLS Method I	125
	7.3 MSLS Method II	130
	7.4 MSLS Method III	132
	7.5 Comparison of the MSLS and GLS Methods	136
	7.6 Concluding Remarks	138

Chapter 8	**Identification of Nonlinear Systems**	141
	8.1 Introduction	141
	8.2 Volterra Series Representation and Identification	141
	8.3 Nonlinear Difference Equations with Linear Parameters	143
	8.4 Nonlinear Difference Equation with Nonlinear Parameters	144
	8.5 Hammerstein Model—GLS Identification	146
	8.6 Hammerstein Model—MSLS Identification	149
	8.7 Concluding Remarks	152
	Index	157
	About the Author	165

List of Figures

1–1	Block Diagram Representation of the System Identification Problem	3
2–1	Single-input and Single-output Discrete System	8
2–2	Multi-input and Multi-output Discrete System	10
2–3	Multivariable Systems with m Inputs and r Outputs, and Weighting Sequences $h_{ij}(k)$	12
2–4	State Variable Equation Representation of Linear Discrete Systems	12
3–1	An n-parameter Linear System	18
3–2	Multi-dependent-variable System	26
4–1	The Basic Configuration for System Identification	39
4–2	The Equivalent Discrete-time System of Figure 4–1	40
4–3	Cross-correlation Identification Scheme	46
4–4	A 15-bit Pseudo-random Binary Sequence (PRBS)	50
4–5	Autocorrelation Function of PRBS	51
4–6	N-stage Shift Register for PRBS Generation	52
4–7	A Digital Computer Identification Scheme	54
4–8	Identification Results of a Second-order Linear System	56
4–9	Learning Model Identification Interpretation of the On-line Least-squares Estimation Algorithm	58
4–10	Multi-input and Multi-output Linear System; $G_{ij}(t)$ is the Weighting Function between Input $u_j(t)$ and Output $w_i(t)$.	60

5–1	A Stable Single-variable Discrete System	69
5–2	Block Diagram Representation of the Error Signal $e(k)$. The Blocks Containing $A(z^{-1})$ and $B(z^{-1})$ Can Be Regarded as Filters.	69
5–3	Systems with Output Noise of Special Type	74
5–4	A Typical Case of $\phi_{ee}(\tau)$ Illustrating the Influence of Model Order n on the Correlation of Residuals for a Second-order System; $n = 2$ is the Correct Model Order	80
5–5	Block Diagram Depiction of the Complete System Identification Algorithm	82
5–6	Real-time Identification Results of a First-order System: (a) with Noise-corrupted Data and $\lambda = 0.90$, (b) with Noise-free Data and $\lambda = 0.90$, (c) with Noise-corrupted Data and $\lambda = 0.99$.	85
5–7	Modeling Configuration of Continuous Systems; $e(t)$ is the Equation Error	88
5–8	Modeling Configuration Using State Variable Filters	90
6–1	Noisy System Model Configuration	98
6–2	Transfer Function Block Diagram of the Noisy System	102
6–3	Block Diagram Showing the Generalized Equation Error	102
6–4	Flow Diagram of the GLS Identification Algorithm	104
6–5	Block Diagram Interpretation of the GLS Algorithm	105
6–6	Rate of Convergence of GLS Algorithm for One Typical Computer Run of the Example System	107
6–7	(a) GLS Algorithm for the Case in Which $v(k)$ is White Noise (b) Minimum-output-error Identification Interpretation of GLS at Convergence	110

6–8	A Different Version of the GLS Algorithm	112
6–9	Rate of Convergence of Three Identification Algorithms for One Typical Computer Run of the Example System	115
6–10	Performance of the System-noise On-line Identification Algorithm	117
6–11	Generation of Instrumental Variables Using an Auxiliary Model	121
7–1	The Three-stage Estimation Scheme of the MSLS Identification Method	126
7–2	Linear System with Additive Output Noise	127
7–3	An Alternative Noise Transfer Function Model Frequently Used in System Identification	131
7–4	Weighting Sequence $\{h_k\}$ of a Third-order Linear System	137
8–1	Nonlinear Dynamic System with Output Additive Noise	142
8–2	Three Types of Nonlinear System Models: (a) Wiener model, (b) Hammerstein model, (c) General model	147
8–3	Hammerstein Model Strusture	147
8–4	Comparison of the Exact and Polynomial-appróximated Saturation Nonlinear Gain	154

List of Tables

4–1	Sum-digit Table of a Module-two Adder	51
5–1	Least-squares Parameter Estimates $\hat{\theta}$ Obtained for Data Length $N = 100$ at Various Noise Levels	75
5–2	Variations of J as a Function of Model Order n for Different Levels of Noise	79
6–1	Computation Time and Memory Requirements of the Bias-correction Algorithms Relative to Those of the GLS Algorithm	115
7–1	Identification Results of a Third-order System, Data Length $N = 300$, Signal/Noise = 1.18, MSE = Mean-square-error, Mean and MSE Computed from 10 Estimation Runs	137
7–2	Identification Results of a Third-order System, Data Length N = 450, Signal/Noise = 1.18, MSE = Mean-square-error, Mean and MSE Computed from 10 Estimation Runs	138
8–1	Parameter Estimates of Example 1 Based on Input-Output data and 10 Computer Runs	153
8–2	Parameter Estimates of Example 2 Based on 450 Input-Output data Points and 10 Computer Runs	153

Preface

System characterization and system identification are very fundamental problems in system engineering practices. System characterization is concerned primarily with setting up mathematical models to represent system input-output relationships. On the other hand, system identification deals with the choice of a specific model for a class of models which is mathematically equivalent to a given physical system.

The application of system identification technology goes beyond the boundaries of engineering and physical sciences. Many other fields of study, such as biological sciences, medicine, and economics, can also benefit by employing system identification methods to establish quantitive models for the systems arising in these areas. Several books on the subject have been published.

A variety of techniques have been devised over the years for system identification. In general the identification techniques are derived from the optimization and estimation theories. The purpose of this book, in contrast to most other books that include a multitude of methods in a single volume, is to focus on the least squares method as a basic solution to the system identification problem. Since the least squares method is a classical method frequently practiced among scientists in various fields, this book can appeal to a large audience. The other motivation for focusing on the least squares method is that other popular identification methods, such as cross-correlation, maximum likelihood, Kalman filtering, instrumental variables and stochastic approximation, can be easily related to the least squares algorithms. Therefore, *System Identification* provides a basis of some degree of integration and unification of many system identification methodologies.

Throughout the book, only systems in open-loop configuration are considered. The basic results can be applied to identify closed-loop systems. In terms of system models, the emphasis is placed on the input-output characterizations (difference, equation and weighting sequence) rather than characterizations by state variables.

The mathematical treatment here is moderate so that the widest possible group of scientists and engineers can participate as readers. However, selected references are also included to allow interested readers to pursue the theoretical developments further. Therefore, the book is suitable as a text for graduate students studying system engineering at the universities.

System Identification

1

Introduction

1.1 The Problem of System Identification

The problem of system identification is generally referred to as the determination of a mathematical model for a system or a process by observing its input-output relationships. It is the purpose of this book to present the basic theory and solution of the system identification problem. Some general references on the subject are listed at the end of the chapter.

The last decade or so has seen tremendous progress in the methodology of system identification. Historically, system identification has been motivated by the need to design better control systems. In most practical systems, such as industrial processes, there is seldom sufficient a priori information about a system and its environment to design an effective control strategy. Very frequently, we are faced with the necessity of experimentally determining some important physical parameters such as heat transfer coefficient, chemical reaction rate, damping factor, and so on. The need for highly accurate system models has been intensified by the development of optimal and adaptive control theories. For example, in adaptive systems design it is often necessary to update the values of some time-varying parameters of a plant and its environment in order to maintain optimal system performance at all times. Other engineering applications for parameter estimation of dynamic systems include communication channel probing, and system and fault testing.

The need for modeling arises in many other disciplines. For example, workers in the field of econometrics have long sought to establish mathematical relationships between endogeneous variables (outputs) and exogenous variables (inputs). More recently, there has been very significant progress toward the application of system identification techniques to physiological and biomedical problems. Very successful models have been obtained for human performance in a man-machine environment, control functions of the pupil and the muscle, metabolism, brain waves, and so on. Thus the subject of system identification is attracting increasing interest from medical and life scientists. Similar modeling applications can be found in such areas as ecology, transportation, and sociology. In addition, the availability of modern estimation theory and sophisticated computational algorithms has contributed to the rapid growth of system identification technology.

1.2 Formulation and Classification of the System Identification Problem

Shown in figure 1-1 is a system with its input and output. The system model we are seeking is the mathematical equation that relates the input to the output at all times. In order to obtain such a model, we are permitted to probe the system with a variety of inputs and observe its responses. The input-output data are then processed to yield the model. On the basis of the degree of a priori knowledge about the system, we can classify the system identification problem into two categories.

1. The complete identification problem: this means that we do not know anything about the basic properties of the system, such as whether it is linear or nonlinear, memoryless or with memory, and so on. Obviously, this is an extremely difficult problem to solve. Usually some kind of assumptions have to be made before any meaningful solution can be attempted. This type of problem is also referred to as a *black box* problem.

2. Partial identification problem: in this category, some basic characteristics of the system, such as linearity, bandwidth, and so on, are assumed to be known. However, we may not know the specific order of the dynamic equation or the values of the associated coefficients. A situation of this kind is also called a *gray box* problem and is, of course, easier to deal with than the black box problem.

Fortunately, the majority of the engineering systems and industrial processes we encounter in practice are of the latter type. In many cases, we know a good deal about the structure of the system, so that it is possible to derive a specific mathematical model of the system dynamics. Consequently, only a set of parameters in the model equation are left to be determined. Thus the modeling problem is reduced to that of *parameter identification*.

Since a majority of system identification problems can be either formulated as or reduced to a parameter identification problem, the treatment of the latter is considered to be of the greatest importance. The present study, as well as most research efforts in the field, have been directed towards this class of problem.

From the viewpoint of system theory, we can precisely determine the unknown parameters in an exact system model equation where the exact measurements of the input-output data are given. In reality, however, the input-output data are corrupted by measurement noise. Furthermore, there are inaccuracies in the model equation as well as random disturbances in the system itself. Therefore, the determination of system parameters is essentially a statistical-estimation problem: we seek to specify a mathematical model that fits the noisy observation data.

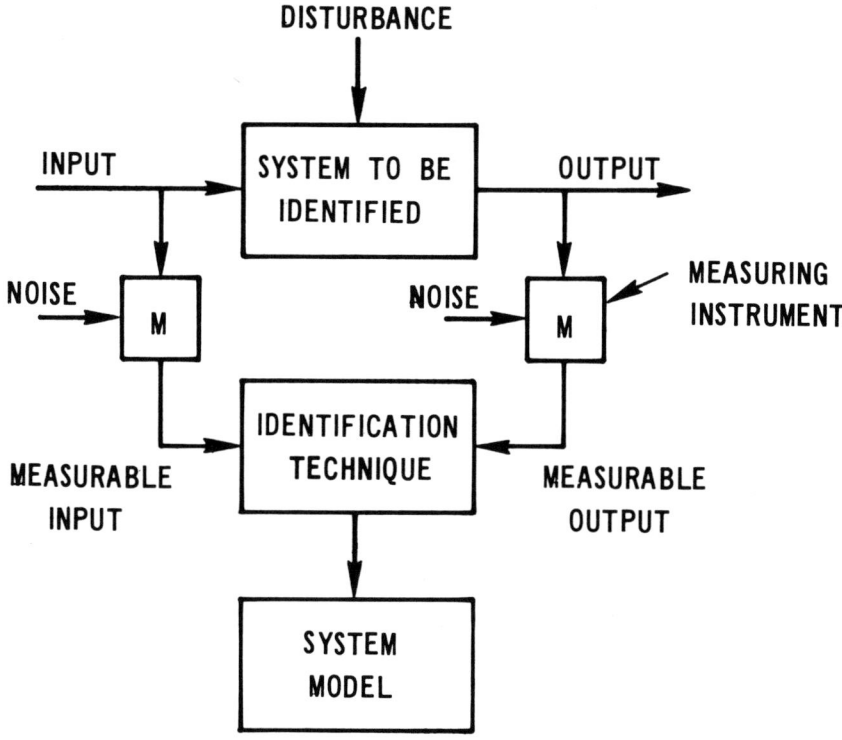

Figure 1–1. Block Diagram Representation of the System Identification Problem

The procedures for carrying out system identification can be divided into the following steps:

1. Specify and parameterize a class of mathematical models that represents the system to be identified.
2. Apply an appropriately chosen test signal to the system and record the input-output data. If the system is in continuous operation and a test signal is not permitted, then we must use the normal operating data for identification.
3. Perform the parameter identification to select the model in the specified class that best fits the statistical data.
4. Perform a validation test to see if the model chosen adequately represents the system with respect to final identification objectives.
5. If the validation test is passed, the procedure ends. Otherwise, another class of models must be selected and steps (2) through (4) performed until a validated model is obtained.

The considerations given in step (1) essentially concern the problem of representation. We can choose a number of possible representations for a given system, including models characterized in the frequency domain or the time domain, in continuous-time or discrete-time. In time domain representation, we have the choices between weighting function (or sequence), differential (or difference) equation, and state variable equation. The choice depends on the identification objective and its associated input-output data.

1.3 Parameter Estimation Methodology

There are a number of well-known parameter estimation techniques that have been successfully applied to the identification problem. They include the methods of maximum likelihood, least squares, cross-correlation, instrumental variable, and stochastic approximation. A variety of other optimization methods have also been proposed.

We have relied primarily on the method of least squares for a number of important reasons. First, the least-squares method is a very classical one with which scientific workers in many disciplines are familiar. Thus it is readily appreciated. Furthermore, the least-squares method is appealing in that it offers conceptual simplicity and applicability to a wide range of situations in which other statistical-estimation theories may be difficult to apply, yet exhibits statistical properties that are as good as those of the maximum likelihood method for most practical situations. In addition, the least-squares identification algorithms can be related easily to many other identification algorithms, making possible a unified treatment of the system identification problem.

As in the case for all estimation techniques, we are seeking to minimize certain appropriately defined error criterion as a means to optimally fit the model to the system data. There are a number of ways by which we can define an error: as the deviation of the parameter estimates from the true values (*parameter error*), as the difference between the output of the system and that of the model in response to the same input (*output error*), or as the discrepancy between the model equation and the measured input and output data (*equation error*). The equation error is the most commonly used.

There are two modes in which identification can be accomplished. One is *off-line identification*, in which a record of input-output data is first observed and the model parameters then estimated based on the entire data record. In *on-line identification*, the parameter estimates are recursively calculated for every data set so that the new data is used to correct and update the existing estimates. Clearly, if the updating process can be made very fast, it becomes possible to obtain parameter estimates of time-varying systems with reasonable accuracy. This capability is called *on-line real-time identification*.

1.4 Organization of the Text

Chapter 2 is devoted to the parametric model representations for dynamic systems. The main emphasis is placed on discrete-time linear system models. Relationships between various forms of representations are pointed out. In chapter 3, we introduce the theory of least squares and the related statistical properties. Relationships to other estimation techniques are established. The application of the least-squares procedures to identify weighting sequence is presented in chapter 4. Concepts of identifiability and optimum test signal design are introduced. In chapter 5, the least-squares procedures for identifying linear difference equation system models are presented for the simple case where the random equation error is white (statistically independent), and the question of model order identification is examined. The case of colored random equation error is examined in chapter 6, and the method of generalized least squares is introduced as the basic identification procedure. In chapter 7, we attempt to solve the same problem by using a number of multistage least-squares procedures. These procedures are computationally simpler than the generalized least-squares techniques discussed in chapter 6. Identification of nonlinear system models is discussed in chapter 8, and particular attention is given to the Volterra series model and the Hammerstein nonlinear model.

References

Åström, K. J., and Eykhoff, P., "System Identification—a Survey," *Automatica*, Vol. 7, pp. 123–162, 1971.

Bekey, G. A., "System identification—an Introduction and a Survey," *Simulation*, pp. 151–166, Oct. 1970.

Cuenod, M., and Sage, A. P., "Comparison of some Methods Used for Process Identification," *Automatica*, Vol. 4, pp. 235–269, 1968.

Eykhoff, P., *System Identification, Parameter and State estimation*, Wiley, London, 1974.

⎯⎯⎯⎯, "Some Fundamental Aspects of Process-Parameter Estimation," *IEEE Transactions on Automatic Control*, Vol. AC-8, pp. 317–357, Oct. 1963.

Graupe, D. *Identification of Systems*, Van Nostrand Reinhold, New York 1972.

Isermann, R.; Baur, R.; Bamberger, W.; Kneppo, P.; and Siebert, H., "Comparison of Six On-line Identification and Parameter Estimation Methods," *Automatica*, Vol. 10, pp. 81–103, Jan. 1974.

Kagiwade, H. H., *System Identification Methods and Applications*, Addison-Wesley, Reading, Mass., 1974.

Lee, R. C. K., *Optimal Estimation, Identification, and Control*, MIT Press, Cambridge, Mass., 1964.

Mendel, J. M., *Discrete Techniques of Parameter Estimation*, Marcel Dekker, New York, 1973.

Phillipson, G. A., *Identification of Distributed Systems*, American Elesevier, New York, 1971.

Saridis, G. N., "Comparison of Six On-line Identification Algorithms," *Automatica*, Vol. 10, pp. 69–79, Jan. 1974.

Sage, A. P., and Melsa, J. L., *System Identification*, Academic Press, New York, 1971.

Zadeh, L. A. "On the Identification Problem," *IRE Transactions on Circuit Theory*, Vol. CT-3, Dec. 1956.

2 Representation of Dynamic Systems

2.1 Introduction

Dynamic system models can be divided into two types: continuous-time and discrete-time. The essential difference between the two is that the signals in the system are continuous in one case and discrete in the other. Since information from continuous signals can be preserved in samples taken at appropriate sampling frequencies, continuous systems can be closely approximated by discrete models. Some general references on system theory are given at the end of the chapter.

The great majority of system identification techniques are digitally oriented as a result of the employment of digital computers. Therefore, discrete system models are more convenient to deal with. The development of the least-squares identification techniques in this book is also based on discrete-system concepts.

In this chapter, we present various forms of representation for the linear discrete dynamic systems. Each form characterizes the system dynamics in a different way. The forms considered are convolution summation, difference equation, and state variable equation. We show that these forms are uniquely related to one another, therefore a linear system can be regarded as completely identified once a particular form of representation is obtained.

2.2 Linear Difference Equation

We will begin by introducing the difference equation representation of a single-variable time-invariant linear discrete system. Referring to the block diagram in figure 2–1, the general nth order difference equation relating the input $u(k)$ and output $y(k)$ is

$$y(k) + a_1 y(k-1) + \ldots + a_n y(k-n) = b_0 u(k) + b_1 u(k-1) + \ldots + b_n u(k-n)$$

or

$$y(k) + \sum_{j=1}^{n} a_j y(k-j) = \sum_{j=0}^{n} b_j u(k-j) \qquad (2.1)$$

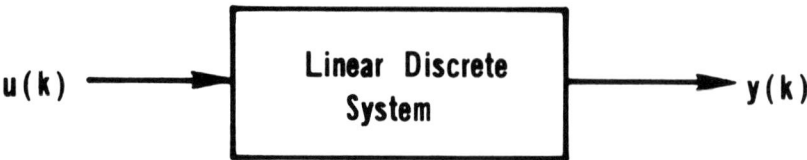

Figure 2-1. Single-input and Single-output Discrete System

where k is the integer time index, and a_j and b_j are the constant coefficients.

If we introduce the shifting operator q defined by

$$q^{-1}y(k) = y(k-1)$$

and the polynomials

$$A(q^{-1}) = 1 + a_1 q^{-1} + a_2 q^{-2} + \ldots + a_n q^{-n}$$
$$B(q^{-1}) = b_0 + b_1 q^{-1} + b_2 q^{-2} + \ldots + b_n q^{-n}, \quad (2.2)$$

equation (2.1) can be written in the form

$$A(q^{-1})y(k) = B(q^{-1})u(k). \quad (2.3)$$

This is the very basic form we will use in system identification.

At this point, it is convenient to indicate the simple relationships between the difference equation and the transfer function representation of a single-variable time invariant system. Applying the Z-transformation on equation (2.1), assuming zero initial conditions $y(k) = u(k) = 0$, $k < 0$, we can obtain the following:

$$(1 + a_1 z^{-1} + \ldots + a_n z^{-n})Y(z) = (b_0 + b_1 z^{-1} + \ldots + b_n z^{-n})U(z),$$

where z is the complex Z-transform variable. The transfer function can then be defined as

$$H(z) = \frac{Y(z)}{U(z)} = \frac{b_0 + b_1 z^{-1} + \ldots + b_n z^{-n}}{1 + a_1 z^{-1} + \ldots + a_n z^{-n}} \quad (2.4)$$

or simply

$$H(z) = \frac{B(z^{-1})}{A(z^{-1})}$$

where the polynomials $A(z^{-1})$ and $B(z^{-1})$ are defined by equation (2.2). We see that the transfer function of a system is directly related to its difference equation (2.3).

The representation of equation (2.1) can be extended to systems that have multiple inputs and outputs. Consider that there are m inputs and r outputs (see figure 2-2), and that we define the vectors $\mathbf{u}(k)$ and $\mathbf{y}(k)$ as

$$\mathbf{u}(k) = \begin{bmatrix} u_1(k) \\ \vdots \\ u_m(k) \end{bmatrix} \quad \mathbf{y}(k) = \begin{bmatrix} y_1(k) \\ \vdots \\ y_r(k) \end{bmatrix}$$

The system can then be represented by the vector difference equation

$$\mathbf{y}(k) + \sum_{j=1}^{n} \mathbf{A}_j \mathbf{y}(k-j) = \sum_{j=0}^{n} \mathbf{B}_j \mathbf{u}(k-j) \tag{2.5}$$

in which $\mathbf{A}_j$ and $\mathbf{B}_j$ are constant coefficient matrices of dimensions $r \times r$ and $r \times m$ respectively. We note here, however, that in general the order of the system in equation (2.5) can be different from n.

It is also possible to put equation (2.5) into the following form:

$$\mathbf{A}(q^{-1})\mathbf{y}(k) = \mathbf{B}(q^{-1})\mathbf{u}(k) \tag{2.6}$$

where $\mathbf{A}(q^{-1})$ and $\mathbf{B}(q^{-1})$ are matrix polynomials in q^{-1} defined by

$$\mathbf{A}(q^{-1}) = \mathbf{I} + \mathbf{A}_1 q^{-1} + \ldots + \mathbf{A}_n q^{-n}$$

$$\mathbf{B}(q^{-1}) = \mathbf{B}_0 + \mathbf{B}_1 q^{-1} + \ldots + \mathbf{B}_n q^{-n}$$

2.3 Weighting Sequence and Convolution

Another basic characterization of the linear discrete system is the weighting sequence. Weighting sequence is defined as the response of a relaxed system to a unit pulse (Kronecker delta function) excitation at $t = 0$. Let the weighting sequence of a causal single-variable system be denoted by

$$\{h(i)\} \quad i = 0, 1, 2, 3, \ldots. \tag{2.7}$$

Then the associated convolution summation for the system input-output relationship (see figure 2.1) is

$$y(k) = \sum_{i=-\infty}^{k} h(k-i)u(i) \tag{2.8}$$

The infinite summation can be simplified in practice when the weight-

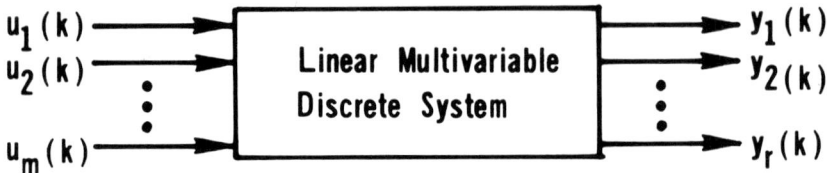

Figure 2–2. Multi-input and Multi-output Discrete System

ing sequence is negligibly small for $i > p$ in equation (2.7). In this case we have the approximation

$$y(k) = \sum_{i=k-p}^{k} h(k-i)u(i) \tag{2.9}$$

It is well known that the Z-transform of the weighting sequence is defined as the transfer function, that is,

$$Z[h(k)] = H(z) \tag{2.10}$$

This equation is very helpful in relating the weighting sequence of a system to its difference equation representation. We will now examine this relationship in some detail.

Recall the transfer function $H(z)$ in equation (2.4). By expanding $H(z)$ by long division, and applying equation (2.10) we can express

$$\frac{b_0 + b_1 z^{-1} + \ldots + b_n z^{-n}}{1 + a_1 z^{-1} + \ldots + a_n z^{-n}} = h_0 + h_1 z^{-1} + h_2 z^{-2} + \ldots \tag{2.11}$$

Rewriting equation (2.5) as

$$b_0 + b_1 z^{-1} + \ldots + b_n z^{-n}$$
$$= (h_0 + h_1 z^{-1} + h_2 z^{-2} + \ldots)(1 + a_1 z^{-1} + \ldots + a_n z^{-n})$$

and comparing the coefficients of the like terms of z^{-1} on both sides, we get the relationships

$$\sum_{m=0}^{i} a_m h(i-m) = \begin{cases} b_i & i = 0, 1, \ldots, n \\ 0 & i > n \end{cases} \tag{2.12}$$

$$a_0 = 1$$

This set of equations ties the weighting sequence $h(i)$ directly to the coefficients a_i and b_i in the difference equation (2.1).

For multivariable systems of m inputs and r outputs, the corresponding representation becomes a weighting matrix $\mathbf{H}(k)$:

$$\mathbf{H}(k) = \begin{bmatrix} h_{11}(k) & \cdots & h_{1m}(k) \\ \vdots & & \vdots \\ h_{r1}(k) & & h_{rm}(k) \end{bmatrix} \qquad (2.13)$$

where $h_{ij}(k)$ is the weighting sequence between the jth input and the ith output. Figure 2–3 depicts such a multivariable system. The corresponding convolution summation is

$$\mathbf{y}(k) = \sum_{i=-\infty}^{k} \mathbf{H}(k-i)\mathbf{u}(i) \qquad (2.14)$$

2.4 State Variable Equation

The single-variable system in figure 2–1 can also be described by the state variable equation

$$\begin{aligned} \mathbf{x}(k+1) &= \mathbf{\Phi}\mathbf{x}(k) + \mathbf{\Gamma} u(k) \\ y(k) &= \mathbf{G}\,\mathbf{x}(k) + \mathbf{D}\,u(k) \end{aligned} \qquad (2.15)$$

in which $\mathbf{x}(k)$ is the $n \times 1$ state vector, and $\mathbf{\Phi}, \mathbf{\Gamma}, \mathbf{G}, \mathbf{D}$ are parameter matrices of dimensions $n \times n$, $n \times 1$, $1 \times n$, 1×1 respectively. A block diagram representation of the state variable equation is shown in figure 2–4.

Here we assume that the state variable system is controllable and observable. (The n-dimensional state equation is controllable if and only if the composite matrix $\mathbf{M}$

$$\mathbf{M} = [\mathbf{\Gamma}, \mathbf{\Phi}\mathbf{\Gamma}, \mathbf{\Phi}^2\mathbf{\Gamma}, \ldots, \mathbf{\Phi}^{n-1}\mathbf{\Gamma}]$$

has rank n. Similarly, the state equation is observable if and only if the composite matrix $\mathbf{V}$

$$\mathbf{V} = [\mathbf{G}^T, \mathbf{\Phi}^T\mathbf{G}^T, (\mathbf{\Phi}^T)^2\mathbf{G}^T, \ldots, (\mathbf{\Phi}^T)^{n-1}\mathbf{G}^T]$$

has rank n.) Thus the weighting sequence and difference equation of the system can be uniquely derived from equation (2.15).

The use of the general form of equation (2.15) to represent a system requires the specification of $n^2 + 2n + 1$ parameters, which is far more than the number of coefficients in the difference equation representation of equation (2.1). Therefore, it is advantageous to consider state equations in canonical forms (minimal parameter representation). Canonical forms can be obtained from an arbitrary form of equation (2.15) by way of

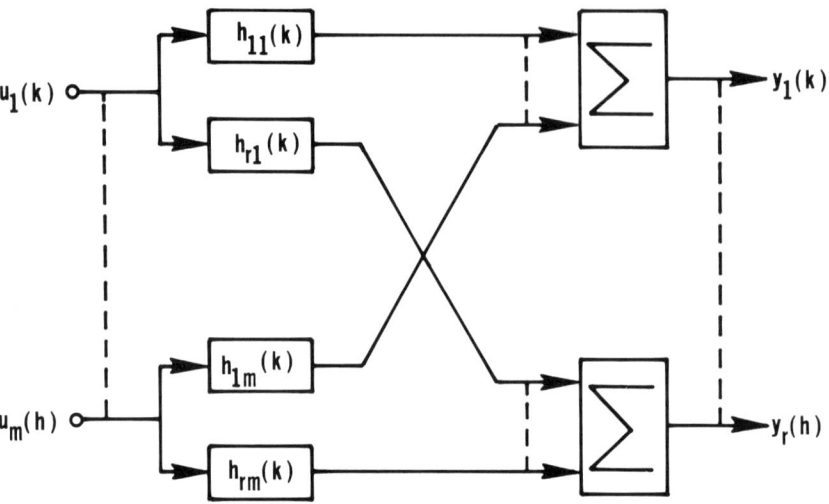

Figure 2–3. Multivariable System with m Inputs and r Outputs, and Weighting Sequences $h_{ij}(k)$

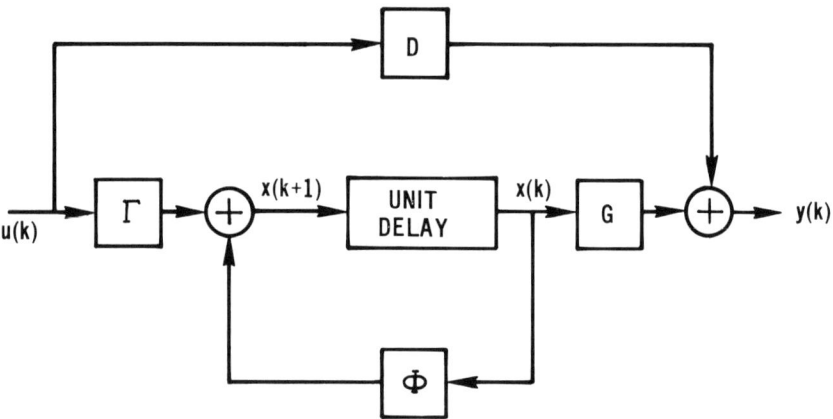

Figure 2–4. State Variable Equation Representation of Linear Discrete Systems

linear transformation. They can also be obtained directly from the system's difference equation (2.1).

One frequently used canonical form state variable equation can be obtained as follows. Transform $\mathbf{x}(k)$ to a new state variable vector $\mathbf{x}^*(k)$ by

$$\mathbf{x}^*(k) = \mathbf{T}\mathbf{x}(k) \qquad (2.16)$$

where **T** is the transformation matrix given by

$$\mathbf{T} = \begin{bmatrix} \mathbf{G} \\ \mathbf{G}\boldsymbol{\Phi} \\ \vdots \\ \mathbf{G}\boldsymbol{\Phi}^{n-1} \end{bmatrix}. \tag{2.17}$$

The resulting new state equation is

$$\mathbf{x}^*(k+1) = \boldsymbol{\Phi}^*(k)\mathbf{x}^*(k) + \boldsymbol{\Gamma}^* u(k)$$
$$y(k) = \mathbf{G}^*\mathbf{x}^*(k) + \mathbf{D}\, u(k) \tag{2.18}$$

in which $\boldsymbol{\Phi}^*$ and $\mathbf{G}^*$ have the simple forms

$$\boldsymbol{\Phi}^* = \left[\begin{array}{c|c} \begin{matrix} 0 & & \\ \vdots & & \mathbf{I} \\ 0 & & \end{matrix} \\ \hline -\phi_n^*, -\phi_{n-1}^*, \ldots, -\phi_1^* \end{array}\right] \tag{2.19}$$

$$\mathbf{G}^* = [1, 0, \ldots, 0] \tag{2.20}$$

and $\boldsymbol{\Gamma}^*$ becomes

$$\boldsymbol{\Gamma}^* = \mathbf{T}\boldsymbol{\Gamma} = \begin{bmatrix} \gamma_1^* \\ \gamma_2^* \\ \vdots \\ \gamma_n^* \end{bmatrix} \tag{2.21}$$

It is noted that the scalar D is unchanged in both equations (2.18) and (2.15).

We see that the total number of parameters in $\boldsymbol{\Phi}^*$, $\boldsymbol{\Gamma}^*$, and D is $2n+1$, which is identical to the number of coefficients in the difference equation (2.1). It can be shown that the parameter relationships between the canonical state equation and the difference equation are

$$\phi_n^* = a_n$$
$$\mathbf{D} = b_0$$

$$\begin{bmatrix} \gamma_1^* \\ \gamma_2^* \\ \vdots \\ \vdots \\ \gamma_n^* \end{bmatrix} = \begin{bmatrix} 1 & & & & \\ a_1 & 1 & & 0 & \\ a_2 & a_1 & 1 & & \\ \vdots & & & & \\ a_{n-1} & \cdots & & a_1 & 1 \end{bmatrix}^{-1} \begin{bmatrix} b_1 - b_0 a_1 \\ b_2 - b_0 a_2 \\ \vdots \\ \vdots \\ b_n - b_0 a_n \end{bmatrix} \quad (2.22)$$

These relationships can be readily established when the state equation (2.18) is converted into the difference equation (2.1).

All other canonical forms can be uniquely specified by $2n + 1$ parameters. All these canonical forms are also uniquely interrelated through linear transformations. For multivariable systems, the form of the state variable equation (2.15) is unchanged except that Γ, $\mathbf{G}$, and $\mathbf{D}$ are now rectangular matrices of appropriate dimensions. However, the canonical representations in this case are more complicated. It can be shown that a canonical form exists only when Φ has distinct eigenvalues or Φ is cyclic (that is, there exists a vector $\mathbf{b}$ such that the vectors $\mathbf{b}$, $\Phi\mathbf{b}$, $\Phi^2\mathbf{b}$, ..., $\Phi^{n-1}\mathbf{b}$ span the n-dimensional space) if the eigenvalues are nondistinct. The number of parameters in the canonical form is $n(m + r) + mr$, with n, m, and r being the dimensions of the vectors $\mathbf{x}$, $\mathbf{u}$, and $\mathbf{y}$ respectively. When matrix Φ has multiple eigenvalues and is not cyclic, the meaning of a minimal parameter representation becomes unclear.

By taking the Z-transform of a relaxed multivariable state equation (2.15), we can obtain the transform relationship

$$\mathbf{Y}(z) = \mathbf{H}(z)\mathbf{U}(z) \quad (2.23)$$

in which $\mathbf{H}(z)$ is the transfer matrix defined by

$$\mathbf{H}(z) = \mathbf{G}(z\mathbf{I} - \Phi)^{-1}\Gamma + \mathbf{D} \quad (2.24)$$

$\mathbf{H}(z)$ is an $r \times m$ matrix

$$\mathbf{H}(z) = [H_{ij}(z)] \quad i = 1, 2, \ldots, r$$
$$j = 1, 2, \ldots, m$$

and $H_{ij}(z)$ is the transfer function between the jth input and the ith output.

For the single-variable case, $r = m = 1$, $\mathbf{H}(z)$ is reduced to a scalar transfer function $H(z)$ having the general form of equation (2.4).

The solution of a multivariable discrete state equation for $t \geq k_0$ is

$$\mathbf{y}(k) = \mathbf{G}\boldsymbol{\Phi}^{k-k_0}\mathbf{x}(k_0) + \sum_{m=k_0}^{k-1} \mathbf{G}\boldsymbol{\Phi}^{k-m-1}\boldsymbol{\Gamma}\mathbf{u}(m) + \mathbf{D}\mathbf{u}(k)$$

Consider that $\mathbf{x}(k_0) = 0$ at $k_0 = -\infty$; then

$$\mathbf{y}(k) = \sum_{m=-\infty}^{k-1} \mathbf{G}\boldsymbol{\Phi}^{k-m-1}\boldsymbol{\Gamma}\mathbf{u}(m) + \mathbf{D}\mathbf{u}(k) \quad (2.25)$$

$$= \sum_{m=-\infty}^{k} \mathbf{H}(k-m)\mathbf{u}(m)$$

in which $\mathbf{H}(k)$ is the weighting sequence matrix defined by

$$\mathbf{H}(k) = \begin{cases} \mathbf{G}\boldsymbol{\Phi}^{k-1}\boldsymbol{\Gamma} & k \geq 1 \\ \mathbf{D} & k = 0 \end{cases} \quad (2.26)$$

These expressions are directly related to expressions given by equations (2.13) and (2.14).

2.5 Concluding Remarks

In this chapter, we have discussed the various types of characterizations for linear discrete systems. We have demonstrated that it is possible to completely represent a given system by any one of the dynamic forms. Furthermore, the interrelationships among these forms are shown to be unique and explicit. Very little attention was given to the frequency domain characteristics, however, since our identification algorithms are based on time domain models only.

References

Chang, C. T., *Introduction to Linear System Theory*, Holt, Rinehart and Winston, New York, 1970.

DeRusso, P. M.; Roy, R. J.; and Close, C. M., *State Variables for Engineers*, Wiley, New York, 1965.

Freeman, H., *Discrete Time Systems*, Wiley, New York, 1965.

3   Least-squares Theory

3.1 Introduction

The purpose of this chapter is to provide a systematic review of the theory of least squares. This review will then provide the basic mathematical foundation for solving the system identification problem in subsequent chapters.

Least-squares theory was first proposed by Karl Gauss for carrying out his work in orbit prediction of planets. Least-squares theory has since become a major tool for parameter estimation from experimental data. Although there are several other estimation methods available, such as maximum likelihood, Baye's method and so on, the least-squares method continues to be the most well known among engineers and scientists. The reason for its popularity is that the method is easier to comprehend than others and does not require a knowledge of mathematical statistics. Furthermore, the least-squares method may provide solutions even in cases when other methods have failed.

Estimates obtained by the least-squares method also have optimal statistical properties: they are consistent, unbiased, and efficient. It also turns out that many estimation algorithms that are used for system identification can be interpreted as least-squares procedures. Therefore, it is possible to unify many identification techniques in the framework of least-squares theory.

3.2 Least-squares Theory

The least-squares technique provides us with a mathematical procedure by which a model can achieve a best fit to experimental data in the sense of minimum-error-squares. Suppose there is a variable y that is related linearly to a set of n variables $\mathbf{x} = (x_1, x_3, \ldots, x_n)$, that is

$$y = \theta_1 x_1 + \theta_2 x_2 + \ldots + \theta_n x_n \qquad (3.1)$$

in which $\boldsymbol{\theta} = (\theta_1, \theta_2, \ldots, \theta_n)$ is a set of constant parameters. We assume here that θ_i are unknown and we wish to estimate their values by observing the variables y and x at different times. A block diagram representation of the problem is shown in figure 3-1.

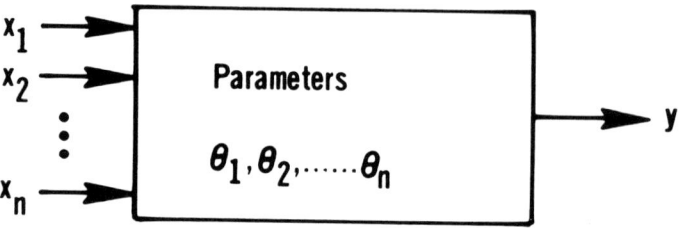

Figure 3-1. An n-parameter Linear System

Let us assume that a sequence of m observations on both y and x has been made at times $t_1, t_2, \ldots, t_m$, and we denote the measured data by $y(i)$ and $x_1(i), x_2(i), \ldots, x_n(i)$, $i = 1, 2, \ldots, m$. Now we can relate these data by the following set of m linear equations:

$$y(i) = \theta_1 x_1(i) + \theta_2 x_2(i) + \ldots + \theta_n x_n(i) \quad i = 1, 2, \ldots, m \quad (3.2)$$

In statistical literature, equation (3.2) is called a *regression function*, and θ_i are the *regression coefficients*.

The system of equations (3.2) can be conveniently arranged into a simple matrix form

$$\mathbf{y} = \mathbf{X}\boldsymbol{\theta} \quad (3.3)$$

where

$$\mathbf{y} = \begin{bmatrix} y(1) \\ y(2) \\ \vdots \\ y(m) \end{bmatrix} \quad \mathbf{X} = \begin{bmatrix} x_1(1) & \ldots & x_n(1) \\ x_1(2) & & x_n(2) \\ \vdots & & \vdots \\ x_1(m) & \ldots & x_n(m) \end{bmatrix} \quad \boldsymbol{\theta} = \begin{bmatrix} \theta_1 \\ \theta_2 \\ \vdots \\ \theta_n \end{bmatrix}$$

To be able to estimate the n parameters θ_i, it is necessary that $m \geq n$. If $m = n$, then we can solve $\boldsymbol{\theta}$ uniquely from equation (3.3) by

$$\hat{\boldsymbol{\theta}} = \mathbf{X}^{-1}\mathbf{y} \quad (3.4)$$

provided that $\mathbf{X}^{-1}$, the inverse of the square matrix $\mathbf{X}$, exists. $\hat{\boldsymbol{\theta}}$ denotes the estimate of $\boldsymbol{\theta}$. However, when $m > n$, it is generally not possible to determine a set of θ_i exactly satsifying all m equations (3.2) because the data may be complicated by random measurement noise, error in the model, or a combination of both. The alternative then is to determine $\boldsymbol{\theta}$ on the basis of least-error-squares.

Define an error vector $\boldsymbol{\epsilon} = (\epsilon_1, \epsilon_2, \ldots, \epsilon_m)^T$ and let

$$\boldsymbol{\epsilon} = \mathbf{y} - \mathbf{X}\boldsymbol{\theta} \tag{3.5}$$

Now we will choose $\hat{\boldsymbol{\theta}}$ in such a way that the criterion J

$$J = \sum_{i=1}^{m} \epsilon_i^2 = \boldsymbol{\epsilon}^T\boldsymbol{\epsilon} \tag{3.6}$$

is minimized. To carry out the minimization, we express

$$J = (\mathbf{y} - \mathbf{X}\boldsymbol{\theta})^T(\mathbf{y} - \mathbf{X}\boldsymbol{\theta})$$
$$= \mathbf{y}^T\mathbf{y} - \boldsymbol{\theta}^T\mathbf{X}^T\mathbf{y} - \mathbf{y}^T\mathbf{X}\boldsymbol{\theta} + \boldsymbol{\theta}^T\mathbf{X}^T\mathbf{X}\boldsymbol{\theta}$$

Differentiate J with respect to $\boldsymbol{\theta}$ and equate the result to zero to determine the conditions on the estimate $\hat{\boldsymbol{\theta}}$ that minimizes J. Thus

$$\left.\frac{\partial J}{\partial \boldsymbol{\theta}}\right|_{\boldsymbol{\theta}=\hat{\boldsymbol{\theta}}} = -2\mathbf{X}^T\mathbf{y} + 2\mathbf{X}^T\mathbf{X}\hat{\boldsymbol{\theta}} = 0$$

This yields

$$\mathbf{X}^T\mathbf{X}\hat{\boldsymbol{\theta}} = \mathbf{X}^T\mathbf{y} \tag{3.7}$$

from which $\hat{\boldsymbol{\theta}}$ can be solved as

$$\hat{\boldsymbol{\theta}} = (\mathbf{X}^T\mathbf{X})^{-1}\mathbf{X}^T\mathbf{y} \tag{3.8}$$

This result is called the *least-squares estimator* (LSE) of $\boldsymbol{\theta}$. Equation (3.7) is referred to as the *normal equation* and $\boldsymbol{\epsilon}$ is called the *residual* in statistical literature.

The above result is derived based on a criterion J that weights every error ϵ_i equally. We often refer to this result as *ordinary least squares*. This formulation can be generalized, however, to allow each error term to be weighted differently. Let $\mathbf{W}$ be the desired weighting matrix. Then the weighted error criterion becomes

$$J_W = \boldsymbol{\epsilon}^T\mathbf{W}\boldsymbol{\epsilon}$$
$$= (\mathbf{y} - \mathbf{X}\boldsymbol{\theta})^T\mathbf{W}(\mathbf{y} - \mathbf{X}\boldsymbol{\theta})$$

Here $\mathbf{W}$ is restricted to being a symmetric positive definite matrix. Minimization of J_W with respect to $\boldsymbol{\theta}$ yields the *weighted least-squares estimator* (WLSE) of $\hat{\boldsymbol{\theta}}_W$:

$$\hat{\boldsymbol{\theta}}_W = (\mathbf{X}^T\mathbf{W}\mathbf{X})^{-1}\mathbf{X}^T\mathbf{W}\mathbf{y} \tag{3.9}$$

It is easy to see that when $\mathbf{W}$ is chosen as an identity matrix $\mathbf{I}$, $\hat{\boldsymbol{\theta}}_W$ is reduced to $\hat{\boldsymbol{\theta}}$.

3.3 Statistical Properties of Least-squares Estimators

In this section we examine the qualities of the least-squares estimators derived above. To facilitate the discussion, we wish to focus on the model equation (3.5) in which the vector $\boldsymbol{\epsilon}$ is included to account for the measurement noise and/or modeling error. Thus we have the noise-disturbed system equation

$$\mathbf{y} = \mathbf{X}\boldsymbol{\theta} + \boldsymbol{\epsilon} \tag{3.10}$$

We assume here that $\boldsymbol{\epsilon}$ is a stationary random vector with zero mean value, that is, $E[\boldsymbol{\epsilon}] = \mathbf{0}$. ($E[\cdot]$ indicates statistical expectation.) Furthermore, $\boldsymbol{\epsilon}$ is uncorrelated with $\mathbf{y}$ and $\mathbf{X}$. Based on these assumptions about $\boldsymbol{\epsilon}$, we wish to know just how good, or how accurate, are the parameter estimates given by equations (3.8) and (3.9).

In general, $\hat{\boldsymbol{\theta}}$ and $\hat{\boldsymbol{\theta}}_W$ are random variables. Their accuracy can be conveniently measured by a number of statistical properties such as bias, error covariance, efficiency, and consistency.

First we show that $\hat{\boldsymbol{\theta}}$ is *unbiased*, meaning that $E[\hat{\boldsymbol{\theta}}] = \boldsymbol{\theta}$. Substituting equation (3.10) into equation (3.8), we have

$$\hat{\boldsymbol{\theta}} = \boldsymbol{\theta} + (\mathbf{X}^T\mathbf{X})^{-1}\mathbf{X}^T\boldsymbol{\epsilon} \tag{3.11}$$

Taking the expectation on both sides of equation (3.11) and applying the property $E[\boldsymbol{\epsilon}] = \mathbf{0}$, we obtain the desired result

$$E[\hat{\boldsymbol{\theta}}] = E[\boldsymbol{\theta}] + E[(\mathbf{X}^T\mathbf{X})^{-1}\mathbf{X}^T]E[\boldsymbol{\epsilon}] = \boldsymbol{\theta} \tag{3.12}$$

Similar proof can be obtained for $E[\hat{\boldsymbol{\theta}}_W] = \mathbf{0}$.

The covariance matrix corresponding to the estimate error $\hat{\boldsymbol{\theta}} - \boldsymbol{\theta}$ is

$$\boldsymbol{\Psi} \stackrel{\Delta}{=} E\{(\hat{\boldsymbol{\theta}} - \boldsymbol{\theta})(\hat{\boldsymbol{\theta}} - \boldsymbol{\theta})^T\}$$

$$= E\{[(\mathbf{X}^T\mathbf{X})^{-1}\mathbf{X}^T\boldsymbol{\epsilon}][(\mathbf{X}^T\mathbf{X})^{-1}\mathbf{X}^T\boldsymbol{\epsilon}]^T\}$$

$$= (\mathbf{X}^T\mathbf{X})^{-1}\mathbf{X}^T E\{\boldsymbol{\epsilon}\boldsymbol{\epsilon}^T\}\mathbf{X}(\mathbf{X}^T\mathbf{X})^{-1}.$$

Define the covariance matrix of the error vector $\boldsymbol{\epsilon}$ to be

$$\mathbf{R} = E[\boldsymbol{\epsilon}\boldsymbol{\epsilon}^T], \tag{3.13}$$

$\boldsymbol{\Psi}$ is reduced to

$$\boldsymbol{\Psi} = (\mathbf{X}^T\mathbf{X})^{-1}\mathbf{X}^T\mathbf{R}\mathbf{X}(\mathbf{X}^T\mathbf{X})^{-1} \tag{3.14}$$

Following the same procedure, we can also show that the error covariance of $\hat{\boldsymbol{\theta}}_W$ is

$$\boldsymbol{\Psi}_W = (\mathbf{X}^T\mathbf{W}\mathbf{X})^{-1}\mathbf{X}^T\mathbf{W}\mathbf{R}\mathbf{W}^T\mathbf{X}(\mathbf{X}^T\mathbf{W}\mathbf{X})^{-1} \tag{3.15}$$

At this point, it is interesting to point out that Ψ_W can be greatly simplified if we let the weighting matrix $\mathbf{W}$ be $\mathbf{W} = \mathbf{R}^{-1}$,

$$\Psi_W \,(\mathbf{W}=\mathbf{R}^{-1}) = (\mathbf{X}^T\mathbf{R}^{-1}\mathbf{X})^{-1} = \Psi_{MV} \tag{3.16}$$

The corresponding estimator $\hat{\boldsymbol{\theta}}_W$ is

$$\hat{\boldsymbol{\theta}}_W(\mathbf{W}=\mathbf{R}^{-1}) = (\mathbf{X}^T\mathbf{R}^{-1}\mathbf{X})^{-1}\mathbf{X}^T\mathbf{R}^{-1}\mathbf{y} = \hat{\boldsymbol{\theta}}_{MV} \tag{3.17}$$

The error covariance Ψ_{MV} in equation (3.16) has a very important property: that is, Ψ_{MV} is a minimum error covariance matrix in the sense that for any other choice of weighting matrices $\mathbf{W}$

$$\Psi_{MV} \leq \Psi_W$$

By definition, a positive definite matrix Ψ_{MV} is less than or equal to Ψ_W if the difference $\Psi_{MV} - \Psi_W$ is non-negative definite. The subscript MV in Ψ_{MV} and $\hat{\boldsymbol{\theta}}_{MV}$ denotes the minimum variance property. The proof of $\Psi_{MV} \leq \Psi_W$ is somewhat involved, and interested readers can see Deutsch (1965).

The estimator $\hat{\boldsymbol{\theta}}_{MV}$ in equation (3.17) is called the *minimum variance estimator*, or *Markov estimator*. Thus we see that $\boldsymbol{\theta}_{MV}$ is the best linear unbiased estimator.

Now let us examine another interesting case. When the noise $\epsilon(i)$, $i = 1, 2, \ldots$, are identically distributed and independent with zero mean and variance σ^2, the covariance $\mathbf{R}$ becomes

$$\mathbf{R} = E[\epsilon\epsilon^T] = \sigma^2\mathbf{I} \tag{3.18}$$

In this case, both Ψ and Ψ_{MV} are identical:

$$\Psi = \Psi_{MV} = \sigma^2(\mathbf{X}^T\mathbf{X})^{-1}. \tag{3.19}$$

This implies that the corresponding *LSE* $\hat{\boldsymbol{\theta}}$ is a minimum variance estimator. $\hat{\boldsymbol{\theta}}$ is called an *efficient estimator*.

Last, we wish to show that the LSE $\hat{\boldsymbol{\theta}}$ is also a consistent estimator. Rewrite the error covariance matrix Ψ in the form of (assume $\mathbf{R} = \sigma^2\mathbf{I}$)

$$\Psi = \sigma^2(\mathbf{X}^T\mathbf{X})^{-1} = \frac{\sigma^2}{m}\left(\frac{1}{m}\mathbf{X}^T\mathbf{X}\right)^{-1}$$

in which m is the number of equations in the vector equation (3.10). Assume that $\lim_{m\to\infty}[(1/m)\,\mathbf{X}^T\mathbf{X}]^{-1} = \Gamma$, where Γ is a nonsingular constant matrix. Then

$$\lim_{m\to\infty} \Psi = \lim_{m\to\infty} \frac{\sigma^2}{m}\left(\frac{1}{m}\mathbf{X}^T\mathbf{X}\right)^{-1} = \mathbf{0}. \tag{3.20}$$

Zero error covariance means that $\hat{\boldsymbol{\theta}} = \boldsymbol{\theta}$ at $m \to \infty$. This convergence property indicates that $\hat{\boldsymbol{\theta}}$ is a consistent estimator.

We have shown that the LSE in the presence of white noise is unbiased, efficient, and consistent. Finally, we wish to note that the LSE $\hat{\boldsymbol{\theta}}$ is also identical to the maximum likelihood estimator (MLE) when the noise ϵ is Gaussian-distributed. This important property is examined in appendix 3A. Thus we see that the least-squares technique does indeed have many advantages.

3.4 Sequential Least-squares Estimation

In this section, we derive a recursive algorithm for the basic least-squares solution in equation (3.8). The need for a recursive solution arises when fresh experimental data are continuously in supply and we wish to improve our parameter estimates by making use of this new information. With a recursive formula, the estimates can be updated step by step without repeatedly computing the matrix solution of equation (3.8), in which the matrix inversion is quite time-consuming. This recursive solution procedure is often referred to as *sequential*, or *on-line estimation*.

Recall that the vector equation (3.3) consists of a set of m equations. Let us introduce m as a subscript to $\mathbf{y}$ and $\mathbf{X}$ in equation (3.3). We have

$$\mathbf{y}_m = \mathbf{X}_m \boldsymbol{\theta} \tag{3.21}$$

Furthermore, denote $\hat{\boldsymbol{\theta}}$ in equation (3.8) as $\hat{\boldsymbol{\theta}}(m)$

$$\hat{\boldsymbol{\theta}}(m) = (\mathbf{X}_m^T \mathbf{X}_m)^{-1} \mathbf{X}_m^T \mathbf{y}_m \tag{3.22}$$

Suppose we have obtained a new equation, the $(m + 1)$th, as

$$y(m+1) = \theta_1 x_1(m+1) + \theta_2 x_2(m+1) + \ldots + \theta_n x_n(m+1).$$

Define

$$\mathbf{x}^T(m+1) = [x_1(m+1), x_2(m+1), \ldots, x_n(m+1)].$$

We then have

$$y(m+1) = \mathbf{x}^T(m+1)\boldsymbol{\theta}. \tag{3.23}$$

Now the system of $m + 1$ equations can be written as

$$\mathbf{y}_{m+1} = \mathbf{X}_{m+1} \boldsymbol{\theta} \tag{3.24}$$

in which

$$\mathbf{y}_{m+1} = \begin{bmatrix} y(1) \\ \vdots \\ y(m) \\ \hline y(m+1) \end{bmatrix} = \begin{bmatrix} \mathbf{y}_m \\ \hline y(m+1) \end{bmatrix} \quad (3.25a)$$

$$\mathbf{X}_{m+1} = \begin{bmatrix} x_1(1) & \cdots & x_n(1) \\ \vdots & & \\ x_1(m) & \cdots & x_n(m) \\ \hline x_1(m+1) & \cdots & x_n(m+1) \end{bmatrix} = \begin{bmatrix} \mathbf{X}_m \\ \hline \mathbf{x}^T(m+1) \end{bmatrix} \quad (3.25b)$$

The new least-squares estimator is

$$\hat{\boldsymbol{\theta}}(m+1) = (\mathbf{X}_{m+1}^T \mathbf{X}_{m+1})^{-1} \mathbf{X}_{m+1}^T \mathbf{y}_{m+1} \quad (3.26)$$

It is apparent that to obtain $\hat{\boldsymbol{\theta}}(m+1)$, we must invert an $n \times n$ matrix. The obvious question here is whether or not we can calculate $\hat{\boldsymbol{\theta}}(m+1)$ by simply updating the previous estimate $\hat{\boldsymbol{\theta}}(m)$ without matrix inversion. The answer is yes, and we derive the updating algorithm below.

To begin, we state the following well-known matrix inversion lemma.

LEMMA (appendix 3B). Let $\mathbf{A}$, $\mathbf{C}$, and $\mathbf{A} + \mathbf{BCD}$ be nonsingular square matrices; then the following matrix identity holds:

$$(\mathbf{A} + \mathbf{BCD})^{-1} = \mathbf{A}^{-1} - \mathbf{A}^{-1}\mathbf{B}(\mathbf{C}^{-1} + \mathbf{D}\mathbf{A}^{-1}\mathbf{B})^{-1}\mathbf{D}\mathbf{A}^{-1} \quad (3.27)$$

Define the matrix $\mathbf{P}(m)$ as

$$\mathbf{P}(m) = (\mathbf{X}_m^T \mathbf{X}_m)^{-1} \quad (3.28)$$

Therefore

$$\mathbf{P}(m+1) = (\mathbf{X}_{m+1}^T \mathbf{X}_{m+1})^{-1}$$

Substituting equation (3.25) and applying the matrix inversion lemma, $\mathbf{P}(m+1)$ can be rewritten as follows:

$$\begin{aligned}\mathbf{P}(m+1) &= [\mathbf{P}(m)^{-1} + \mathbf{x}(m+1)\mathbf{x}^T(m+1)]^{-1} \\ &= \mathbf{P}(m) - \mathbf{P}(m)\mathbf{x}(m+1)[1 + \mathbf{x}^T(m+1)\mathbf{P}(m)\mathbf{x}(m+1)]^{-1} \\ &\quad \cdot \mathbf{x}^T(m+1)\mathbf{P}(m). \end{aligned} \quad (3.29)$$

In view of equation (3.26), we can see that

$$\hat{\boldsymbol{\theta}}(m+1) = \mathbf{P}(m+1)[\mathbf{X}_m^T \mathbf{y}_m + \mathbf{x}(m+1)y(m+1)]$$
$$= \mathbf{P}(m)\mathbf{X}_m^T \mathbf{y}_m - \mathbf{P}(m)\mathbf{x}(m+1)[1 + \mathbf{x}^T(m+1)$$
$$\cdot \mathbf{P}(m)\mathbf{x}(m+1)]^{-1} \cdot \mathbf{x}^T(m+1)\mathbf{P}(m)\mathbf{X}_m^T \mathbf{y}_m$$
$$+ \mathbf{P}(m)\mathbf{x}(m+1)y(m+1) - \mathbf{P}(m)\mathbf{x}(m+1)$$
$$\cdot [1 + \mathbf{x}^T(m+1)\mathbf{P}(m)\mathbf{x}(m+1)]^{-1}$$
$$\cdot \mathbf{x}^T(m+1)\mathbf{P}(m)\mathbf{x}(m+1)y(m+1)$$

We can rearrange the last two terms into the form of

$$\mathbf{P}(m)\mathbf{x}(m+1)[1 + \mathbf{x}^T(m+1)\mathbf{P}(m)\mathbf{x}(m+1)]^{-1}$$
$$\cdot [1 + \mathbf{x}^T(m+1)\mathbf{P}(m)\mathbf{x}(m+1) - \mathbf{x}^T(m+1)\mathbf{P}(m)\mathbf{x}(m+1)]y(m+1)$$
$$= \mathbf{P}(m)\mathbf{x}(m+1)[1 + \mathbf{x}^T(m+1)\mathbf{P}(m)\mathbf{x}(m+1)]^{-1}y(m+1)$$

But we recognize from equations (3.22) and (3.28) that

$$\hat{\boldsymbol{\theta}}(m) = \mathbf{P}(m)\mathbf{X}_m^T \mathbf{y}_m$$

Thus $\hat{\boldsymbol{\theta}}(m+1)$ can finally be simplified to the form

$$\hat{\boldsymbol{\theta}}(m+1) = \hat{\boldsymbol{\theta}}(m) + \mathbf{P}(m)\mathbf{x}(m+1)[1 + \mathbf{x}^T(m+1)\mathbf{P}(m)\mathbf{x}(m+1)]^{-1}$$
$$\cdot [y(m+1) - \mathbf{x}^T(m+1)\hat{\boldsymbol{\theta}}(m)] \quad (3.30)$$

The result above simply shows that the new estimate is given by the old estimate plus a correction term. The matrix $\mathbf{P}(m)$ in the correction term can be updated by the recursive formula in equation (3.29). It is clear that in both formulas we have completely eliminated the necessity of matrix inversion (we note that the term $[1 + \mathbf{x}^T(m+1)\mathbf{P}(m)\mathbf{x}(m+1)]$ is a scalar) and therefore that the computational efficiency is drastically improved for updating the estimate $\hat{\boldsymbol{\theta}}$.

The recursive equation (3.30) has a very strong intuitive appeal. We notice that the correction term is proportional to the quantity $y(m+1) - \mathbf{x}^T(m+1)\hat{\boldsymbol{\theta}}(m)$, which represents the error of fitting the previous estimate $\hat{\boldsymbol{\theta}}(m)$ to the new data $y(m+1)$ and $\mathbf{x}^T(m+1)$. The vector $\mathbf{P}(m)\mathbf{x}(m+1)$ $\cdot [1 + \mathbf{x}^T(m+1)\mathbf{P}(m)\mathbf{x}(m+1)]^{-1}$ determines how the fitting error is weighted in the correction of $\hat{\boldsymbol{\theta}}(m)$. Another interesting fact is that $\mathbf{P}(m)$ can be related to the error covariance matrix $\boldsymbol{\Psi}$ defined by equation (3.19). It shows that $\mathbf{P}(m) = \boldsymbol{\Psi}(m)/\sigma^2$, which means that $\mathbf{P}(m)$ is a direct measure of the error covariance at each m. As we have shown in equation (3.20), $\mathbf{P}(m) = \mathbf{0}$ at the limit $m \to \infty$.

We have shown in this section that sequential least-squares estimation can be easily carried out by the following recursive algorithm:

$$\hat{\theta}(m+1) = \hat{\theta}(m) + \gamma(m+1)\mathbf{P}(m)\mathbf{x}(m+1)[y(m+1)$$
$$- \mathbf{x}^T(m+1)\hat{\theta}(m)] \qquad (3.31a)$$
$$\mathbf{P}(m+1) = \mathbf{P}(m) - \gamma(m+1)\mathbf{P}(m)\mathbf{x}(m+1)\mathbf{x}^T(m+1)\mathbf{P}(m) \qquad (3.31b)$$

where

$$\gamma(m+1) = 1/[1 + \mathbf{x}^T(m+1)\mathbf{P}(m)\mathbf{x}(m+1)].$$

Therefore, by starting with an initial estimate $\hat{\theta}(0)$ and the corresponding $\mathbf{P}(0)$, we can sequentially update $\hat{\theta}$ while new observations are continuously obtained.

Finally, we wish to comment on the selection of the initial values of $\hat{\theta}$ and $\mathbf{P}$ for starting the algorithm. Two practical approaches are discussed below.

1. Take the first m data points and solve $\hat{\theta}(m)$ and $\mathbf{P}(m)$ directly from

$$\hat{\theta}(m) = (\mathbf{X}_m^T\mathbf{X}_m)^{-1}\mathbf{X}_m^T y_m$$
$$\mathbf{P}(m) = (\mathbf{X}_m^T\mathbf{X}_m)^{-1} \qquad (3.32)$$

This then allows the algorithm to iterate from $m+1$ data points onward.

2. Set $\hat{\theta}(0)$ arbitrarily and $\mathbf{P}(0) = \alpha I$ where α is a very large positive scalar, and I is the identity matrix. The algorithm iterates from $\hat{\theta}(1)$ and $\mathbf{P}(1)$. We now wish to show that at mth iteration, the values of $\hat{\theta}(m)$ and $\mathbf{P}(m)$ approach that in equation (3.32) when α approaches infinity.

Starting from $\mathbf{P}(0)$, and iterating m times using equation (3.31b), we get a $\mathbf{P}(m)$ such that

$$\mathbf{P}(m) = [\mathbf{P}(0)^{-1} + \mathbf{X}_m^T\mathbf{X}_m]^{-1} \qquad (3.33)$$

The corresponding $\hat{\theta}$ is

$$\hat{\theta}(m) = \mathbf{P}(m)[\mathbf{X}_m^T y_m + \mathbf{P}(0)^{-1}\hat{\theta}(0)].$$

To make equation (3.33) agree with equation (3.32), we simply force $\mathbf{P}(0)^{-1}$ to zero. This can easily be done by letting $\alpha \to \infty$ since

$$\lim_{\alpha \to \infty} \mathbf{p}(0)^{-1} = \lim_{\alpha \to \infty} \frac{1}{\alpha} \mathbf{I} = 0.$$

3.5 Multi-dependent-variable System

The least-squares theory presented in the previous sections can be generalized to multi-dependent-variable systems. Shown in figure 3–2 is a system with p dependent variables y_i, $i = 1, 2, \ldots, p$. The system equations are

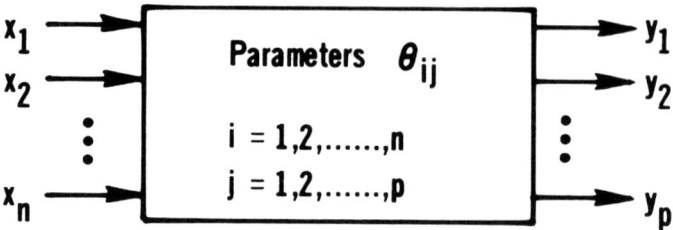

Figure 3-2. Multi-dependent-variable System

$$y_1 = \theta_{11}x_1 + \theta_{21}x_2 + \ldots + \theta_{n1}x_n$$
$$\vdots \qquad (3.34)$$
$$y_p = \theta_{1p}x_1 + \theta_{2p}x_2 + \ldots + \theta_{np}x_n$$

The corresponding vector equation is

$$\mathbf{y} = \mathbf{x}\Theta \qquad (3.35)$$

where $\mathbf{y} = [y_1, y_2, \ldots, y_p]$ and $\mathbf{x} = [x_1, x_2, \ldots, x_n]$

$$\Theta = \begin{bmatrix} \theta_{11} & & \theta_{1p} \\ \vdots & & \vdots \\ \theta_{n1} & \ldots & \theta_{np} \end{bmatrix} = [\boldsymbol{\theta}_1, \boldsymbol{\theta}_2, \ldots, \boldsymbol{\theta}_p] \qquad (3.36)$$

Now suppose a set of m measurements are obtained from the system at times $t_1, t_2, \ldots, t_m$. We can then arrange the equations in the following form:

$$\mathbf{Y} = \mathbf{X}\Theta \qquad (3.37)$$

where

$$\mathbf{Y} = \begin{bmatrix} \mathbf{y}(1) \\ \mathbf{y}(2) \\ \vdots \\ \mathbf{y}(m) \end{bmatrix} \quad \mathbf{X} = \begin{bmatrix} \mathbf{x}(1) \\ \mathbf{x}(2) \\ \vdots \\ \mathbf{x}(m) \end{bmatrix} \qquad (3.38)$$

and $\mathbf{x}(i)$, $\mathbf{y}(i)$ are vectors of $\mathbf{x}$ and $\mathbf{y}$ measured at t_i.

The least-squares estimate of the parameter matrix $\mathbf{\Theta}$ can be solved by minimizing the $p \times p$ matrix function $\mathbf{J}$

$$\mathbf{J} = (\mathbf{Y} - \mathbf{X}\mathbf{\Theta})^T(\mathbf{Y} - \mathbf{X}\mathbf{\Theta}) \tag{3.39}$$

with respect to $\mathbf{\Theta}$. The form of the solution is similar to that of the single-dependent-variable case, that is

$$\hat{\mathbf{\Theta}} = (\mathbf{X}^T\mathbf{X})^{-1}\mathbf{X}^T\mathbf{Y} \tag{3.40}$$

When comparing with the solution (3.8), we see that the ith column of $\hat{\mathbf{\Theta}}$, $\hat{\mathbf{\Theta}}_i$ is

$$\hat{\boldsymbol{\theta}}_i = (\mathbf{X}^T\mathbf{X})^{-1}\mathbf{X}^T\mathbf{y}_i \tag{3.41}$$

in which $\mathbf{y}_i$ is the ith column of $\mathbf{Y}$. This implies that the parameter $\hat{\theta}_{1i}$, $\hat{\theta}_{2i}$, ..., $\hat{\theta}_{ni}$ corresponding to the dependent variable y_i can be estimated separately from other parameters in the system. Consequently, the sequential least-squares algorithms introduced in section 3.4 are directly applicable to each set of parameters $(\hat{\theta}_{1i}, \hat{\theta}_{2i}, \ldots, \hat{\theta}_{ni})$. Thus we see that a multi-dependent-variable problem is basically not different from the single-dependent-variable problem.

3.6 Recursive Estimation for Increasing Parameter Numbers

In this section, we present an algorithm that allows us to recursively compute the least-squares estimates as the number of parameters increases. In many regression analysis problems, the exact number of parameters needed to fit the data is usually unknown ahead of time. Thus we would have to try to make the fit with different parameter numbers in the model equation. But this would require us to repeatedly solve the matrix equation (3.8), a computationally wasteful procedure. A preferable approach would be to make use of the computations already completed in the p-parameter problem to compute the estimates of the q-parameter $(q > p)$ problem. In this way, only a $(q - p) \times (q - p)$ matrix inversion is needed. We now present this recursive algorithm.

Let the parameter vector θ be partitioned into two parts as follows:

$$\boldsymbol{\theta} = [\theta_1, \theta_2, \ldots, \theta_p \vdots \theta_{p+1}, \ldots, \theta_q]^T$$

$$= \begin{bmatrix} \boldsymbol{\theta}_{(1)} \\ \text{----} \\ \boldsymbol{\theta}_{(2)} \end{bmatrix} \tag{3.42}$$

The corresponding **X** matrix is then

$$\mathbf{X} = \begin{bmatrix} x_1(1), & \ldots, & x_p(1) & \vdots & x_{p+1}(1), & \ldots, & x_1(1) \\ \vdots & & & \vdots & & & \\ x_1(m), & \ldots, & x_p(m) & \vdots & x_{p+1}(m), & \ldots, & x_q(m) \end{bmatrix}$$

$$= [\mathbf{X}_1 \vdots \mathbf{X}_2] \qquad (3.43)$$

Thus, according to equation (3.7), we have

$$\begin{bmatrix} \mathbf{X}_1^T\mathbf{X}_1 & \vdots & \mathbf{X}_1^T\mathbf{X}_2 \\ \text{------} & & \text{------} \\ \mathbf{X}_2^T\mathbf{X}_1 & \vdots & \mathbf{X}_2^T\mathbf{X}_2 \end{bmatrix} \begin{bmatrix} \boldsymbol{\theta}_{(1)} \\ \text{---} \\ \boldsymbol{\theta}_{(2)} \end{bmatrix} = \begin{bmatrix} \mathbf{X}_1^T\mathbf{y} \\ \text{---} \\ \mathbf{X}_2^T\mathbf{y} \end{bmatrix}$$

This yields

$$\mathbf{X}_1^T\mathbf{X}_1\hat{\boldsymbol{\theta}}_{(1)} + \mathbf{X}_1^T\mathbf{X}_2\hat{\boldsymbol{\theta}}_{(2)} = \mathbf{X}_1^T\mathbf{y}$$

$$\mathbf{X}_2^T\mathbf{X}_1\hat{\boldsymbol{\theta}}_{(1)} + \mathbf{X}_2^T\mathbf{X}_2\hat{\boldsymbol{\theta}}_{(2)} = \mathbf{X}_2^T\mathbf{y}$$

Solve these two equations for $\hat{\boldsymbol{\theta}}_{(1)}$ and $\hat{\boldsymbol{\theta}}_{(2)}$

$$\hat{\boldsymbol{\theta}}_{(1)} = \hat{\hat{\boldsymbol{\theta}}}_{(1)} - \mathbf{A}\mathbf{X}_2^T(\mathbf{y} - \mathbf{X}_1\hat{\hat{\boldsymbol{\theta}}}_{(1)})$$

$$\hat{\boldsymbol{\theta}}_{(2)} = \mathbf{B}\mathbf{X}_2^T(\mathbf{y} - \mathbf{X}_1\hat{\boldsymbol{\theta}}_{(1)}) \qquad (3.44)$$

in which

$$\mathbf{A} = (\mathbf{X}_1^T\mathbf{X}_1)^{-1}\mathbf{X}_1^T\mathbf{X}_2\mathbf{B}$$

$$\mathbf{B} = [\mathbf{X}_2^T\mathbf{X}_2 - \mathbf{X}_2^T\mathbf{X}_1(\mathbf{X}_1^T\mathbf{X}_1)^{-1}\mathbf{X}_1^T\mathbf{X}_2]^{-1}$$

$$\hat{\hat{\boldsymbol{\theta}}}_{(1)} = (\mathbf{X}_1^T\mathbf{X}_1)^{-1}\mathbf{X}_1^T\mathbf{y}$$

$\hat{\hat{\boldsymbol{\theta}}}_{(1)}$ is the least squares estimate of the q-parameter system.

Equation (3.44) shows how we can compute the new q-parameter estimates $(\hat{\boldsymbol{\theta}}_{(1)}, \hat{\boldsymbol{\theta}}_{(2)})$ from the available knowledge of lower-order p-parameter estimates $\hat{\hat{\boldsymbol{\theta}}}_{(1)}$. The term $(\mathbf{y} - \mathbf{X}_1\hat{\boldsymbol{\theta}}_{(1)})$ can be interpreted as being the residual vector of the p-parameter model. We note that matrix inversion in **B** involves a matrix of order $(q - p)$ only. To enable the algorithm in equation (3.44) to compute the parameters of the next-higher-order model, we need to compute the matrix inverse $(\mathbf{X}^T\mathbf{X})^{-1}$ as required in **A** and **B**. We use the property of partitioned matrix inverse. Let R be an $n \times n$ nonsingular matrix partitioned as

$$\mathbf{R} = \begin{bmatrix} \mathbf{E} & \mathbf{F} \\ \mathbf{G} & \mathbf{H} \end{bmatrix}$$

where $\mathbf{E}$ is $n_1 \times n_2$, $\mathbf{F}$ is $n_1 \times n_2$, $\mathbf{G}$ is $n_2 \times n_1$, $\mathbf{H}$ is $n_2 \times n_2$. Suppose that $\mathbf{E}$ and $\mathbf{D} = \mathbf{H} - \mathbf{GE}^{-1}\mathbf{F}$ are nonsingular. Then

$$\mathbf{R}^{-1} = \begin{bmatrix} \mathbf{E}^{-1}(\mathbf{I} - \mathbf{FD}^{-1}\mathbf{GE}^{-1}) & -\mathbf{E}^{-1}\mathbf{FD}^{-1} \\ -\mathbf{D}^{-1}\mathbf{GE}^{-1} & \mathbf{D}^{-1} \end{bmatrix}$$

We then can show that

$$(\mathbf{X}^T\mathbf{X})^{-1} = \begin{bmatrix} \mathbf{X}_1^T\mathbf{X}_1 & \mathbf{X}_1^T\mathbf{X}_2 \\ \mathbf{X}_2^T\mathbf{X}_1 & \mathbf{X}_2^T\mathbf{X}_2 \end{bmatrix}^{-1}$$

$$= \begin{bmatrix} \mathbf{C} - \mathbf{A}\mathbf{X}_2^T\mathbf{X}_1\mathbf{C} & -\mathbf{A} \\ -\mathbf{A}^T & \mathbf{B} \end{bmatrix} \quad (3.45)$$

where $\mathbf{C} = (\mathbf{X}_1^T\mathbf{X}_1)^{-1}$. Equations (3.44) (3.45) constitute the complete recursive algorithm. This algorithm will be applied in chapter 5.

3.7 Real-time Least-squares Algorithm

The sequential least-squares algorithm presented in section 3.4 was developed based on a policy of equal weighting for all the measured data as the process evolves. The reason for using equal weighting was that the parameters were essentially constant throughout the period of estimation, so that the most recent data was as good as older data for providing information about the unknown parameter values. However, when this algorithm is applied to a situation where the parameters to be estimated are time-varying, the estimates can easily become erratic and do not bear a close resemblance to the true time variation of the parameter values.

A sequential algorithm that is able to closely track time-varying parameters is called a *real-time algorithm*. In this section, we present such a real-time algorithm: it is a simple modification of the least-squares algorithm in which an exponential weighting scheme is used to place heavier emphasis on the more recent data. As a result, the parameter tracking capability is greatly increased.

Consider the error function

$$J_m = \sum_{i=1}^{m} \lambda^{m-i} \epsilon^2(i) \quad 0 < \lambda < 1 \qquad (3.46)$$

in which the later squared errors are given more weight than earlier ones. Define a weighted residual vector ϵ_m by

$$\epsilon_m = [\sqrt{\lambda^{m-1}} \epsilon(1), \sqrt{\lambda^{m-2}} \epsilon(2), \ldots, \epsilon(m)]^T$$

and the corresponding weighted system equation (see equation (3.5))

$$\epsilon_m = y_m - X_m \theta$$

Then

$$J_m = \epsilon_m^T \epsilon_m \qquad (3.47)$$

and the corresponding least-squares parameter estimate is

$$\hat{\theta}(m) = (X_m^T X_m)^{-1} X_m^T y_m \qquad (3.48)$$

For $m + 1$ data set, we can define

$$J_{m+1} = \sum_{i=1}^{m+1} \lambda^{m+1-i} \epsilon^2(i)$$

Then

$$J_{m+1} = \lambda \sum_{i=1}^{m} \lambda^{m-i} \epsilon^2(i) + \epsilon^2(m+1)$$

$$= \lambda \epsilon_m^T \epsilon_m + \epsilon^2(m+1)$$

$$= \begin{bmatrix} \sqrt{\lambda} \epsilon_m \\ \hdashline \epsilon(m+1) \end{bmatrix} \begin{bmatrix} \sqrt{\lambda} \epsilon_m \\ \hdashline \epsilon(m+1) \end{bmatrix} \qquad (3.49)$$

Since

$$\sqrt{\lambda} \epsilon_m = \sqrt{\lambda} y_m - \sqrt{\lambda} X_m \theta$$

and

$$\epsilon(m+1) = y(m+1) - x(m+1)^T \theta$$

the least-squares parameter estimate from J_{m+1} is

$$\hat{\theta}(m+1) = \left\{ \left[\begin{array}{c} \sqrt{\lambda}\mathbf{X}_m \\ \hline \mathbf{x}^T(m+1) \end{array} \right]^T \left[\begin{array}{c} \sqrt{\lambda}\mathbf{X}_m \\ \hline \mathbf{x}^T(m+1) \end{array} \right] \right\}^{-1}$$

$$\times \left[\begin{array}{c} \sqrt{\lambda}\mathbf{X}_m \\ \hline \mathbf{x}(m+1) \end{array} \right]^T \left[\begin{array}{c} \sqrt{\lambda}\mathbf{y}_m \\ \hline y(m+1) \end{array} \right] \quad (3.50)$$

We see that equation (3.50) is identical in form to equation (3.26); therefore we can follow the same procedures outlined earlier to derive the recursive algorithm for $\hat{\theta}(m+1)$ in equation (3.51). This time, we define $P(m)$ as

$$\mathbf{P}(m) = \frac{1}{\lambda}(\mathbf{X}_m^T\mathbf{X}_m)^{-1}$$

Then we can show that

$$\hat{\theta}(m+1) = \hat{\theta}(m) + \gamma(m+1)\mathbf{P}(m)\mathbf{x}(m+1)$$
$$\cdot [y(m+1) - \mathbf{x}^T(m+1)\hat{\theta}(m)] \quad (3.51)$$

$$\mathbf{P}(m+1) = \frac{1}{\lambda}[\mathbf{P}(m) - \gamma(m+1)\mathbf{P}(m)\mathbf{x}(m+1)\mathbf{x}^T(m+1)\mathbf{P}(m)]$$

$$\gamma(m+1) = 1/[1 + \mathbf{x}^T(m+1)\mathbf{P}(m)\mathbf{x}(m+1)]$$

We note that the above equation is reducible to equation (3.31) when $\lambda = 1$. We observe that the smaller the λ, the heavier are the weights assigned to the more recent data. This implies that the algorithm is more capable of tracking the parameter variations. However, the estimates may also fluctuate more because of noise disturbance. Therefore, γ is often chosen experimentally for a given task. We shall say more about this in chapter 5.

3.8 Concluding Remarks

In this chapter, we have reviewed the basic theories of least-squares parameter estimation. The fundamental matrix solutions for both the ordinary and weighted least-squares estimators were derived. We have also examined the statistical properties of these estimators under random observation noise conditions. It has been shown that the least-squares estimators have the properties of being unbiased, efficient, and consistent.

Sequential estimation algorithms have also been derived, and they will be useful later for on-line system identification applications.

References

Åström, K.J., *Lectures on the Identification Problem—the Least Squares Method*, Report 6086, Lund Institute of Technology, Division of Automatic Control, 1968.

Deutsch, R., *Estimation Theory*, Prentice Hall, Englewood Cliffs, N.J., 1965.

Eykhoff, P., *System Identification*, Wiley, New York, 1974.

Mendel, J.M., *Discrete Techniques of Parameter Estimation*, Marcel Dekker, New York, 1973.

Sage, A.P., and Melsa, J.L., *Estimation Theory with Application to Communications and Control*, McGraw-Hill, New York, 1971.

Appendix 3A: Maximum Likelihood Estimation

The method of maximum likelihood is a very old standard technique in estimation theory. The basic concept is simple and can be stated as follows:

Assume that ϵ is a discrete white random process and we know the form of its probability density function $p(\epsilon; \theta)$ where θ is some unknown parameter. Suppose we take a set of m independent samples, $\epsilon_1, \epsilon_2, \ldots, \epsilon_m$. We then ask what the best estimate is of the parameter θ based on these samples. To find the answer to this question, we proceed as follows. The estimate θ is chosen in such a way that the samples ϵ_i are most likely to occur during these independent measurements. To do this we must define a likelihood function and then maximize it with respect to θ.

The likelihood function L is usually defined as the joint probability density function of ϵ_i. Since ϵ_i is serially uncorrelated, we can write

$$L(\epsilon_1, \ldots, \epsilon_m; \theta) = \prod_{i=1}^{m} p(\epsilon_i; \theta) \tag{3A.1}$$

Equation (3A.1) indicates that the likelihood function is simply the product of the individual probability functions of the samples.

Since $\log L$ is a monotonic function and attains its maximum when L is greatest, maximizing L with respect to θ is then equivalent to solving the following equation:

$$\frac{\partial}{\partial \theta} \log L \big|_{\theta=\hat{\theta}} = 0 \tag{3A.2}$$

which is called the *likelihood equation*.

Now let us consider the most important special case, that ϵ_i is Gaussian-distributed with zero mean and variance σ^2. Furthermore, we consider the linear vector equation $\epsilon = y - X\theta$, which has been defined in equation (3.5). Knowing that the Gaussian probability density function for ϵ_i has the form

$$p(\epsilon_i, \theta) = \frac{1}{(2\pi\sigma^2)^{1/2}} \operatorname{Exp}\left(-\frac{1}{2\sigma^2} \epsilon_i^2\right)$$

we can show that

$$\log L(\theta) = \log \prod_{i=1}^{m} p(\epsilon_i, \theta) = \log\left\{\left(\frac{1}{2\pi\sigma^2}\right)^{m/2} \operatorname{Exp}\left(-\frac{1}{2\sigma^2} \epsilon^T \epsilon\right)\right\}$$

$$= -\frac{1}{2\sigma^2}(\mathbf{y} - \mathbf{X}\boldsymbol{\theta})(\mathbf{y} - \mathbf{X}\boldsymbol{\theta}) - \frac{m}{2}\log(2\pi\sigma^2)$$

$$= -\frac{1}{2\sigma^2}(\mathbf{y} - \mathbf{X}\boldsymbol{\theta})^T(\mathbf{y} - \mathbf{X}\boldsymbol{\theta}) + \text{constant}. \tag{3A.3}$$

Minimizing log L with respect to $\boldsymbol{\theta}$ yields the likelihood equation

$$\frac{\partial}{\partial \boldsymbol{\theta}}(\mathbf{y} - \mathbf{X}\boldsymbol{\theta})^T(\mathbf{y} - \mathbf{X}\boldsymbol{\theta})|_{\boldsymbol{\theta} = \hat{\boldsymbol{\theta}}} = \mathbf{0} \tag{3A.4}$$

The solution of $\hat{\boldsymbol{\theta}}$ from equation (3A.4) is identical to the least-squares solution given by equation (3.8). Thus we have established the well-known principle that the maximum likelihood estimator for Gaussian noise is equivalent to the least-squares estimator.

Appendix 3B: Matrix Inversion Lemma

We wish to give a simple proof to the matrix inversion lemma stated below. The proof directly follows that given by Åström (1968).

LEMMA. Let **A**, **C**, and **A** + **BCD** be nonsingular square matrices, then the matrix identity

$$(A + BCD)^{-1} = A^{-1} - A^{-1}B(C^{-1} + DA^{-1}B)^{-1}DA^{-1} \qquad (3B.1)$$

holds.

Proof. Premultiply both sides of (3B.1) by **A** + **BCD**,

$$I = (A + BCD)[A^{-1} - A^{-1}B(C^{-1} + DA^{-1}B)^{-1}DA^{-1}]. \qquad (3B.2)$$

The objective now is to prove the identity (3B.2). In other words, we wish to show that the right hand side of (3B.2) can be reduced to an identity matrix. By direct manipulation, we get

$$[A + BCD][A^{-1} - A^{-1}B(C^{-1} + DA^{-1}B)^{-1}DA^{-1}]$$

$$= I + BCDA^{-1} - B(C^{-1} + DA^{-1}B)^{-1}DA^{-1}$$

$$\quad - BCDA^{-1}B(C^{-1} + DA^{-1}B)^{-1}DA^{-1}$$

$$= I + BCDA^{-1} - B[I + CDA^{-1}B][C^{-1} + DA^{-1}B]^{-1}DA^{-1}$$

$$= I + BCDA^{-1} - BC[C^{-1} + DA^{-1}B][C^{-1} + DA^{-1}B]^{-1}DA^{-1}$$

$$= I + BCDA^{-1} - BCDA^{-1}$$

$$= I$$

Reference

Åström, K.J., *Lectures on the Identification Problem—the Least Squares Method*, Report 6086, Lund Institute of Technology, Division of Automatic Control, 1968.

4 Weighting Function Identification

4.1 Introduction

In this chapter, we present the least-squares technique for identifying the weighting functions of linear dynamic systems. Along with frequency response representation, weighting function is basically a nonparametric model. On the other hand, representations such as state variable equations and transfer functions are classified as parametric models. Historically, weighting functions have formed the nucleus of system analysis and design. Although state variable models have gained prominence in modern control theory, the weighting function approach is still considered very effective and desirable.

Estimating weighting function under noisy conditions was the engineer's earliest attempts at systems identification. Many estimation techniques have since been developed, for example, the methods of matched filter (Turin 1957), cross-correlation (Anderson, Buland, and Cooper 1959), and least squares (Levin 1960). When the noise level is sufficiently low, the simpler methods based on the concepts of error-correction (Nagumo and Noda 1967; Mendel 1968) and numerical deconvolution (Graupe 1972), are also very effective for identifying weighting functions.

There are a number of obvious advantages of the weighting function model from the viewpoint of system identification. First, the determination of the weighting function requires less a priori knowledge than do the parametric models. For example, the troublesome questions of the system order and time delay in parametric models do not arise. Second, this model can be identified more satisfactorily in the presence of noise. These points are demonstrated in chapter 5. Because of these advantages, the weighting function model is often preferred as the basis for exploratory studies. For these reasons we choose to discuss this identification problem first.

In this chapter, we concentrate on the least-squares identification method. This method is digitally oriented, and therefore only sample points of a continuous weighting function can be estimated. To implement the identification algorithm, system input and output signals must be digitized, and the continuous system approximated by a discrete model represented by the convolution summation. The samples of weighting function are then identified by estimating the weighting sequence in the discrete system model.

We also introduce the important concepts of identifiability and optimum test signal design. The discussions in this chapter will lay the foundations for identification of parametric models.

4.2 The Identification Problem

In figure 1–1, we have depicted the general configuration for the system identification problem. It was pointed out that there are various sources of disturbances and measurement noises that corrupt the input and output signals of the system. An exact analysis of these random noises and their effects on the accuracy of identification can be highly complicated. However, since we are dealing with a linear system, it is possible to lump all the noise effects into a single additive noise source at the systems output. Shown in figure 4–1 is such a simplified configuration. This is also the very basic configuration for which our identification problem is formulated.

In figure 4–1, the input $u(t)$ is often a specifically designed test signal or a control signal under which the system is operating normally. Thus we can say that $u(t)$ is directly observable. The true system output $w(t)$ is assumed to be inaccessible for measurement. Instead, we can only observe the noisy output $y(t)$. The noise $v(t)$ is a stationary random process that is uncorrelated to both $u(t)$ and $w(t)$.

Let the system be linear, causal, and time-invariant. Denote the system's unknown weighting function by $g(t)$. In view of figure 4–1, we can describe the system by the convolution integral

$$w(t) = \int_{-\infty}^{t} g(t - \lambda) u(\lambda) d\lambda \tag{4.1}$$

where

$$y(t) = w(t) + v(t) \tag{4.2}$$

Suppose $u(t)$ and $y(t)$ are sampled periodically with a sampling period of T seconds. Furthermore, if T is sufficiently small, we can take the piecewise constant approximations

$$u(t) \approx u(kT), \quad g(t) \approx g(kT) \text{ for } kT \leq t < (k + 1)T$$

Then, at $t = kT$, equation (4.1) can be closely approximated by

$$w(kT) = \sum_{i=-\infty}^{k} Tg(kT - iT) u(iT) \tag{4.3}$$

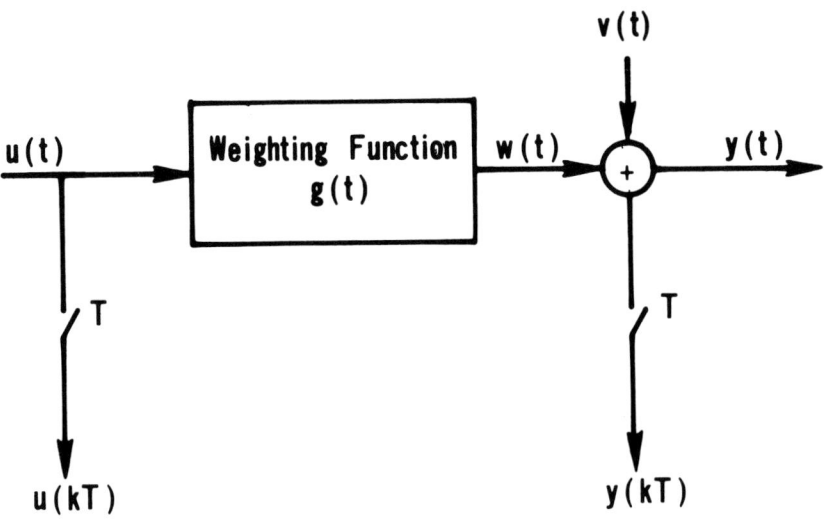

Figure 4-1. The Basic Configuration for System Identification

Define an equivalent weighting sequence $h(kT)$ as

$$h(kT) = Tg(kT) \qquad (4.4)$$

Then

$$w(kT) = \sum_{i=-\infty}^{k} h(kT - iT)u(iT) \qquad (4.5)$$

which is equivalent to the convolution summation of a discrete system shown in figure 4-2.

Let us assume that the system is stable, and its settling time is finite so that $g(t) \approx 0$ for some $t > pT$, where p is an integer. Then combining equations (4.2) and (4.5), we get the overall system equation

$$y(kT) = \sum_{i=k-p}^{k} h(kT - iT)u(iT) + v(kT) \qquad (4.6)$$

Now we can state the identification problem as follows.

Given the input-output sampled sequences $\{u(kT)\}\{y(kT)\}$, and the integer p, estimate the weighting sequence $\{h(kT)\}$, $k = 0, 1, 2, \ldots, p$. The desired weighting function samples $g(kT)$ are then determinable from $g(kT) = h(kT)/T$.

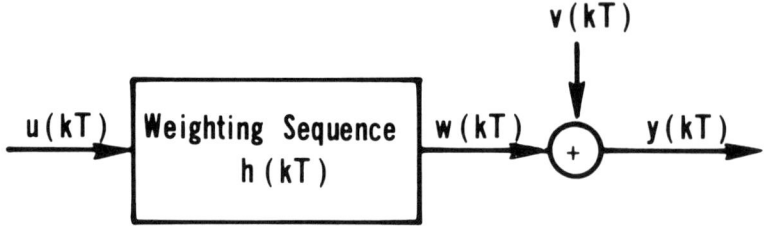

Figure 4–2. The Equivalent Discrete-time System of Figure 4–1

4.3 Least-squares Estimation

In this section we develop the complete solution of the identification problem stated above. To start, we make the following additional assumptions:

1. The input $u(t)$ is a part of continuing driving function of the system. Starting from $t = 0$, we observe the input sequence $\{u(kT)\}$ for $0 \le k \le m + p$ where $m > p$.
2. The output sequence $\{y(kT)\}$ for $p \le k \le m + p$ is also observed.
3. The noise $v(kT)$ is a random sequence with zero mean, $E[v(kT)] = 0$.

Using the observed input-output data in equation (4.6), we can set up a set of $m + 1$ equations written in the vector form

$$\mathbf{y} = \mathbf{U}\mathbf{h} + \mathbf{v} \tag{4.7}$$

where

$$\mathbf{y} = [y(p), y(p+1), \ldots, y(p+m)]^T$$
$$\mathbf{v} = [v(p), v(p+1), \ldots, v(p+m)]^T$$
$$\mathbf{h} = [h(0), h(1), \ldots, h(p)]^T$$

$$\tag{4.8}$$

$$\mathbf{U} = \begin{bmatrix} u(p) & u(p-1) & \cdots & u(0) \\ u(p+1) & u(p) & & u(1) \\ \vdots & & & \\ u(p+m) & u(p+m-1) & \cdots & u(m) \end{bmatrix}$$

We note that in the above matrices the sampling time T is dropped from the arguments of u, y and v for simplicity.

Since $m > p$, we can now estimate the unknown parameter vector $\mathbf{h}$ by the method of least squares introduced in chapter 3. Based on the procedures stated in section 3.2 of chapter 3, we define the error vector

$$\mathbf{v} = \mathbf{y} - \mathbf{Uh}, \tag{4.9}$$

and then minimize the error criterion J

$$J = \mathbf{v}^T\mathbf{v} = (\mathbf{y} - \mathbf{Uh})^T(\mathbf{y} - \mathbf{Uh})$$

with respect to $\mathbf{h}$. This leads to the normal equation

$$(\mathbf{U}^T\mathbf{U})\hat{\mathbf{h}} = \mathbf{U}^T\mathbf{y} \tag{4.10}$$

from which we obtain the least-squares estimator $\hat{\mathbf{h}}$

$$\hat{\mathbf{h}} = (\mathbf{U}^T\mathbf{U})^{-1}\mathbf{U}^T\mathbf{y} \tag{4.11}$$

By setting $\hat{g}(kT) = \hat{h}(kT)/T$, $k = 0, 1, \ldots, p$, the samples of the weighting function $g(t)$ are then identified.

Now we wish to examine the accuracy of the estimates $\hat{\mathbf{h}}$. To do this, we must assume that the system indeed satisfies the equation

$$w(kT) = \sum_{i=k-p}^{k} h(kT - iT) u(iT) \tag{4.12}$$

and that a set of true values of $\{h(kT)\}$ exists. We can then measure the accuracy of $\hat{\mathbf{h}}$ against the assumed true $\mathbf{h}$. Since $\hat{\mathbf{h}}$ is random, we will examine the statistical properties of $\hat{\mathbf{h}}$.

1. $\hat{h}$ is an unbiased estimate of $\mathbf{h}$ since

$$E[\hat{\mathbf{h}}] = E[(\mathbf{U}^T\mathbf{U})^{-1}\mathbf{U}^T(\mathbf{Uh} + \mathbf{v})]$$
$$= E[\mathbf{h}] + E[(\mathbf{U}^T\mathbf{U})^{-1}\mathbf{U}^T]E[\mathbf{v}] = \mathbf{h}$$

2. In view of equation (3.14), the error covariance of $\hat{\mathbf{h}}$ is

$$\mathbf{\Psi} = E[(\hat{\mathbf{h}} - \mathbf{h})(\hat{\mathbf{h}} - \mathbf{h})^T] = (\mathbf{U}^T\mathbf{U})^{-1}\mathbf{U}^T\mathbf{R}\mathbf{U}(\mathbf{U}^T\mathbf{U})^{-1} \tag{4.13}$$

in which $\mathbf{R}$ is the covariance matrix of $\mathbf{v}$, $\mathbf{R} = E[\mathbf{v}\mathbf{v}^T]$. If $v(k)$ is an identically distributed independent random variable with variance σ_v^2, then $\mathbf{R} = \sigma_v^2\mathbf{I}$ and

$$\mathbf{\Psi} = \sigma_v^2(\mathbf{U}^T\mathbf{U})^{-1} \tag{4.14}$$

The estimate $\hat{\mathbf{h}}$ is now called the *minimum variance (efficient) estimator*. An unbiased estimate of σ_v^2 can be determined from the minimized error criterion $J(\hat{\mathbf{h}})$ as (Levin 1960)

$$\hat{\sigma}_v^2 = \frac{1}{m+1}J(\hat{\mathbf{h}}) = \frac{1}{m+1}(\mathbf{y} - \mathbf{U}\hat{\mathbf{h}})^T(\mathbf{y} - \mathbf{U}\hat{\mathbf{h}})$$

$$= \frac{1}{m+1}[\mathbf{y}^T\mathbf{y} - \mathbf{y}^T\mathbf{U}\hat{\mathbf{h}}]$$

3. If the noise $v(k)$ is both independent and Gaussian, then $\hat{\mathbf{h}}$ is also the maximum likelihood estimate of $\mathbf{h}$.

4. Assume that, as the length of observation m approaches infinity, the following limit exists:

$$\lim_{m\to\infty} \frac{1}{m} \mathbf{U}^T\mathbf{U} = \mathbf{\Phi} \tag{4.15}$$

where $\mathbf{\Phi}$ is a nonsingular constant matrix. Then we can follow equation (3.20) to show that $\hat{\mathbf{h}}$ converges to $\mathbf{h}$ in the manner that

$$\lim_{m\to\infty} \mathbf{\Psi} = \lim_{m\to\infty} \frac{\sigma_v^2}{m}\left(\frac{1}{m}\mathbf{U}^T\mathbf{U}\right)^{-1} = 0 \tag{4.16}$$

Hence, $\hat{\mathbf{h}}$ is a consistent estimator of $\mathbf{h}$.

The principle that the parameter such as $\mathbf{h}$ in a system can be consistently estimated is often referred to as the *identifiability* of a system. Moreover, the input signal $u(kT)$ that satisfies the property stated in equation (4.15) is called *persistently exciting* of order $(p+1)$. (We call an input sequence $\{u(k)\}$ persistently exciting of order n if

$$\lim_{N\to\infty} \frac{1}{N}\sum_{k=1}^{N} u(k) = \bar{u}$$

and

$$\lim_{N\to\infty} \frac{1}{N}\sum_{k=1}^{N}[u(k) - \bar{u}][u(k+\tau) - \bar{u}] = \phi_{uu}(\tau)$$

exist, and

$$\begin{bmatrix} \phi_{uu}(0) & \cdots & \phi_{uu}(n-1) \\ & \vdots & \\ \phi_{uu}(n-1) & \cdots & \phi_{uu}(0) \end{bmatrix}$$

is positive definite.) The assumption that the weighting function of a linear system has finite settling time implies that the system is stable. Hence, a linear system is identifiable if the system is stable and the input test signal is persistently exciting.

Finally, we wish to remark on the selection of the settling time parameter p. Before we proceed with the solution (4.11), we must choose a priori a value for p. We recognize that, for a fixed T, the larger the p the better the sequence $\{h(kT)/T\}$ approximates $g(kT)$. However, a large p increases the computational difficulties associated with high order matrix inversion in equation (4.11). Therefore we should pick a reasonably small p according to the condition $g(t > pT) \approx 0$. In many cases, we may want to start with a small p, and then repeat the solution with progressively increasing p values until a satisfactory fit has been achieved.

4.4 Relationship to Cross-correlation Identification

In this section, we show that the least-squares solution (4.11) can be expressed in terms of empirical auto- and cross-correlation functions of the system's input and output. This formulation allows us to relate the least-squares solution to the well-known cross-correlation identification method (Anderson, Buland, and Cooper 1959).

First, let us rewrite equation (4.11) as

$$\hat{\mathbf{h}} = (\mathbf{U}^T\mathbf{U})^{-1}\mathbf{U}^T\mathbf{y} = \left(\frac{1}{m+1}\mathbf{U}^T\mathbf{U}\right)^{-1}\left(\frac{1}{m+1}\mathbf{U}^T\mathbf{y}\right)$$

or

$$\hat{\mathbf{h}} = \mathbf{\Phi}^{-1}\mathbf{\Gamma} \tag{4.17}$$

where

$$\mathbf{\Phi} = \frac{1}{m+1}\mathbf{U}^T\mathbf{U}, \quad \mathbf{\Gamma} = \frac{1}{m+1}\mathbf{U}^T\mathbf{y} \tag{4.18}$$

In view of equation (4.8) we can show that

$$\mathbf{U}^T\mathbf{U} = \begin{bmatrix} \sum_{j=p}^{p+m} u^2(j) & \sum_{j=p}^{p+m} u(j)u(j-1) & \cdots & \sum_{j=p}^{p+m} u(j)u(j-p) \\ & \sum_{j=p-1}^{p+m-1} u^2(j) & \cdots & \sum_{j=p-1}^{p+m-1} u(j)u(j-p+1) \\ \text{(symmetric)} & & \ddots & \vdots \\ & & & \sum_{j=0}^{m} u^2(j) \end{bmatrix}$$

$$\mathbf{U}^T\mathbf{y} = \begin{bmatrix} \sum_{j=p}^{p+m} u(j)y(j) \\ \sum_{j=p}^{p+m} u(j-1)y(j) \\ \vdots \\ \sum_{j=p}^{p+m} u(j-p)y(j) \end{bmatrix} \qquad (4.19)$$

Assume that the input-output sequences $\{u(k)\}$ and $\{y(k)\}$ are stationary ergodic random processes. Now define the following m-point empirical auto- and cross-correlation functions

$$\phi_{uu}(k) = \frac{1}{m+1} \sum_{j=i}^{i+m} u(j)u(j-k)$$

$$\phi_{uy}(k) = \frac{1}{m+1} \sum_{j=i}^{i+m} u(j-k)y(j) \qquad (4.20)$$

in which i is an arbitrary positive integer indicating when in the time sequences the correlations start to be computed. For sufficiently large m, $\phi_{uu}(k)$ and $\phi_{uy}(k)$ are nearly invariant with respect to i if the input-output time sequences are stationary. Based on these assumptions, the matrices $\mathbf{\Phi}$ and $\mathbf{\Gamma}$ in equation (4.17) can then be expressed as

$$\mathbf{\Phi} = \begin{bmatrix} \phi_{uu}(0) & \phi_{uu}(1) & \phi_{uu}(2) & \cdots & \phi_{uu}(p) \\ & \phi_{uu}(0) & \phi_{uu}(1) & \cdots & \phi_{uu}(p-1) \\ \text{(Sym.)} & & & \ddots & \vdots \\ & & & & \phi_{uu}(0) \end{bmatrix} \qquad (4.21)$$

$$\mathbf{\Gamma} = \begin{bmatrix} \phi_{uy}(0) \\ \phi_{uy}(1) \\ \phi_{uy}(2) \\ \vdots \\ \phi_{uy}(p) \end{bmatrix} \qquad (4.22)$$

The significance of this result is that for sufficiently long data length, estimation of **h** can be accomplished by first computing the various correlation functions, and then computing equation (4.17).

Let us now consider the interesting case that $\{u(k)\}$ is a sequence of stationary independent variables. Then the autocorrelation function $\phi_{uu}(k)$ becomes a Kronecker delta function as $m \to \infty$,

$$\lim_{m \to \infty} \phi_{uu}(k) = G\delta(k)$$

in which G is the mean square value of the random input $\{u(k)\}$, or $G = \phi_{uu}(0)$. The important consequence of this is that $\boldsymbol{\Phi}$ becomes a diagonal matrix so that $\boldsymbol{\Phi}^{-1} = (1/G)\mathbf{I}$. This immediately simplifies the solution (4.17) to $\hat{\mathbf{h}} = (1/G)\boldsymbol{\Gamma}$. It is noted that the elements of **h** are directly proportional to the cross-correlation functions

$$\hat{h}(i) = \frac{1}{G} \phi_{uy}(i) \quad i = 0, 1, 2, \ldots, p. \tag{4.23}$$

This result is the sampled data analog of cross-correlation identification (Anderson, Buland, and Cooper 1959). The entire identification scheme is shown in figure 4-3. This scheme shows that system impulse response samples can be determined directly from the cross-correlation functions of the system input and output sampled data.

There is another interesting result that is worth pointing out here. In view of equation (4.17) we can write

$$\boldsymbol{\Phi}\hat{\mathbf{h}} = \boldsymbol{\Gamma} \tag{4.24}$$

Based on the definitions of $\boldsymbol{\Phi}$ and $\boldsymbol{\Gamma}$, and the symmetry property of autocorrelation functions

$$\phi_{uu}(i) = \phi_{uu}(-i),$$

the components in equation (4.24) can be written in the following form

$$\sum_{i=0}^{p} \hat{h}(i) \phi_{uu}(i-j) = \phi_{uy}(j) \quad j = 0, 1, 2, \ldots, p. \tag{4.25}$$

This equation is the sampled data analog of the well known *Wiener-Hopf equation* (Wiener 1949).

We have shown in this section that the cross-correlation identification has the properties of least-squares estimation. The point of interest here is that when the system input is a white random process, the matrix $\boldsymbol{\Phi}$ in the least-squares solution becomes a diagonal matrix that permits the decoupled solution of $\hat{h}(i)$. Finally, relationship to Wiener-Hopf equation has also been established for the least-squares solution.

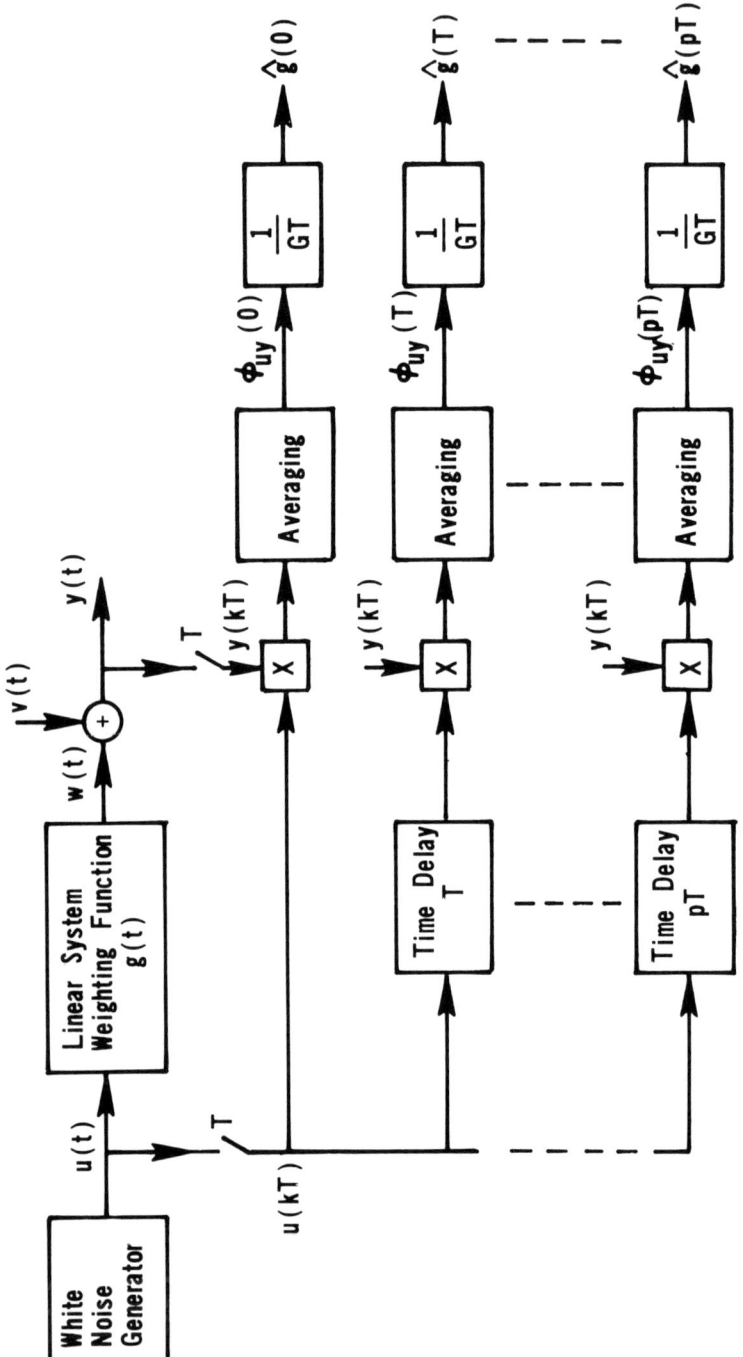

Figure 4-3. Cross-correlation Identification Scheme

4.5 Optimum Input Signal

The unique feature of the cross-correlation identification presented above is that the input test signal is a white random process. Consequently, this technique allows each component of the weighting sequence $h(i)$ to be estimated individually as indicated in equation (4.23). Moreover, the estimation errors of all $\hat{h}(i)$ are independent from one another. The importance of this property is that the estimation error of $\hat{\mathbf{h}}$ is the smallest for a class of inputs of which the white random process is a member.

Intuitively we can appreciate the fact that different types of input signals produce $\hat{\mathbf{h}}$ of varying degrees of accuracy. From the standpoint of parameter estimation, we seek to synthesize the particular input signal that gives the most accurate estimates. This special input is referred to as the *optimum input signal* (Goodwin, Murdock, and Paine 1973; Mehra 1974). In this section we provide a more quantitative analysis of the white random process as an optimum input signal.

Recall that the error covariance matrix $\boldsymbol{\Psi}$ of $\hat{\mathbf{h}}$ for the case in which the disturbing noise $v(k)$ is white was given in equation (4.14) as

$$\boldsymbol{\Psi} = \sigma_v^2 (\mathbf{U}^T \mathbf{U})^{-1} \tag{4.26}$$

Improving the accuracy of $\mathbf{h}$ implies in some sense decreasing the magnitude of $\boldsymbol{\Psi}$. Since $\boldsymbol{\Psi}$ is proportional to the inverse of $\mathbf{U}^T \mathbf{U}$, it is apparent that the magnitude of $\boldsymbol{\Psi}$ can be reduced if the magnitude of $u(k)$ is increased. (Physically, strong $u(k)$ implies high signal-to-noise ratio at the system output. Theoretically, exact parameter values can be determined if the signal-to-noise ratio goes to infinity.) In practice, however, the magnitude of $u(k)$ is limited by power considerations and nonlinearities that exist in generating the signal. Therefore, we must consider the class of inputs that are constrained in either power or magnitude. Once an appropriate constraint is imposed, we then ask what wave form $u(k)$ should take so as to minimize $\boldsymbol{\Psi}$.

Let us now specifically consider the constraint that the mean square value of $u(k)$ is fixed as

$$\frac{1}{m+1} \sum_{k=0}^{m} u^2(k) = \phi_{uu}(0) = G \tag{4.27}$$

Define the correlation coefficients γ_i as

$$\gamma_i = \frac{\phi_{uu}(i)}{\phi_{uu}(0)} \quad i = 1, 2, \ldots, p.$$

Then

$$\mathbf{U}^T\mathbf{U} = (m + 1)\phi_{uu}(0) \begin{bmatrix} 1 & \gamma_1 & \gamma_2 & \cdots & \gamma_p \\ & 1 & \gamma_1 & & \vdots \\ & & \ddots & & \vdots \\ & & & & 1 \end{bmatrix}$$

$$= (m + 1)\phi_{uu}(0)\mathbf{R} \tag{4.28}$$

It is recognized that the diagonal elements of $\mathbf{\Psi}$ are the individual error variances of $\hat{h}(i)$. Therefore, the aim is to choose $u(k)$ so as to minimize the diagonal elements in $(\mathbf{U}^T\mathbf{U})^{-1}$. Since the quantity $(m + 1)\phi_{uu}(0)$ in equation (4.28) is fixed, the only possibility is to optimize the matrix $\mathbf{R}$ by selecting appropriate values of γ_i. Levin (1960) has shown that the minimum diagonal values of $\mathbf{R}^{-1}$ achievable are unity provided that $\gamma_i = 0$ for $i = 1, 2, \ldots, p$. This means that $\mathbf{R}$ is an identity matrix. Hence

$$\mathbf{U}^T\mathbf{U} = (m + 1)\phi_{uu}(0)\mathbf{I}$$

and

$$\mathbf{\Psi} = \frac{1}{(m + 1)\phi_{uu}(0)}\mathbf{I} \tag{4.29}$$

The conditions that $u(k)$ must satisfy are

$$\phi_{uu}(0) = G$$
$$\phi_{uu}(i) = 0 \quad 0 < i \le p \tag{4.30}$$

This means that the optimal input signal $u(k)$ is uncorrelated (or white) over a range of p sampling intervals. Obviously a band-limited white noise sequence qualifies as an optimal input signal.

The physical implication of the optimal input signal characteristics of equation (4.30) is that $u(k)$ has a wide power spectrum, so that the system is excited by a wide spectrum of frequencies. Therefore, the output contains maximum information about the dynamic modes of the system. Consequently, parameter estimates obtained under these conditions are the most accurate. Ideally, the impulse function, which has a flat power spectrum, is the best input signal. However, it is impossible to generate in practice.

The above discussions illustrate what is involved in the design of the optimum input signals for system identification. When different constraints other than those of equation (4.27) are imposed, different signal characteristics would result. The synthesis of these signals can be quite complicated, and we do not pursue this problem any further.

4.6 Pseudo-random Binary Sequence

In the last two sections, we have shown that white random noise is a desirable input signal for weighting sequence estimation. It has been demonstrated that, under white noise excitation, the matrix $U^T U$ in the least-squares solution is diagonal, for which matrix inversion $(U^T U)^{-1}$ is avoided. Furthermore, the cross-correlations between input and output data provide directly the weighting function samples. This means that deconvolution is unnecessary. Estimators under such decoupling conditions have been shown to be most accurate in the sense of showing the least error variance. In this section we discuss a practical way of generating such a test signal for the purpose of system identification.

The practical white noise signal that we can easily generate by a digital circuit, or digital computer, is the so-called pseudo-random binary sequence (PRBS). The PRBS is a periodic sequence that takes on only two values. The times at which transition can occur are multiples of a specified time interval Δt, and the state for any succeeding interval is nearly independent of the state in any preceding interval.

An example of such a signal is shown in figure 4–4. This signal has a periodic autocorrelation function shown in figure 4–5. We see that the autocorrelation function closely approximates the delta function of an ideal white noise. The approximation can be adjusted by changing N and Δt. The bias in the PRBS autocorrelation function can be minimized if N is large ($N \geq 100$, for instance) or compensated by adding a corrective d.c. signal to the PRBS. The periodic structure of PRBS provides the advantage of simpler implementation.

PRBS can be easily generated by a logically controlled shift register circuit as shown in figure 4–6. The binary output represents the two states of the PRBS. The modulo-two adder in the feedback path has the property shown in table 4–1. The length of the binary sequence in a period generated by a given shift register generator depends on both the feedback connections and on the initial loading of the shift register. The largest possible period for an n-stage shift register is $N = (2^n - 1)$. The output sequence that has a period of $N = (2^n - 1)$ is called a *maximum-length sequence*. These types of signals and their properties are extensively discussed by Davies (1970). A PRBS digital computer simulation program is given in appendix 4A.

Since PRBS is periodic, its autocorrelation function is also necessarily periodic. In order to maintain proper operation of the cross-correlation identification technique of equation (4.23), it is necessary to choose a PRBS period τ ($\tau = N\Delta t$) large enough such that $\phi_{uy}(\tau) \approx 0$. In other words, τ is necessarily greater than the settling time t_s of the weighting function of the system. When τ is smaller than t_s, overlapping of weighting

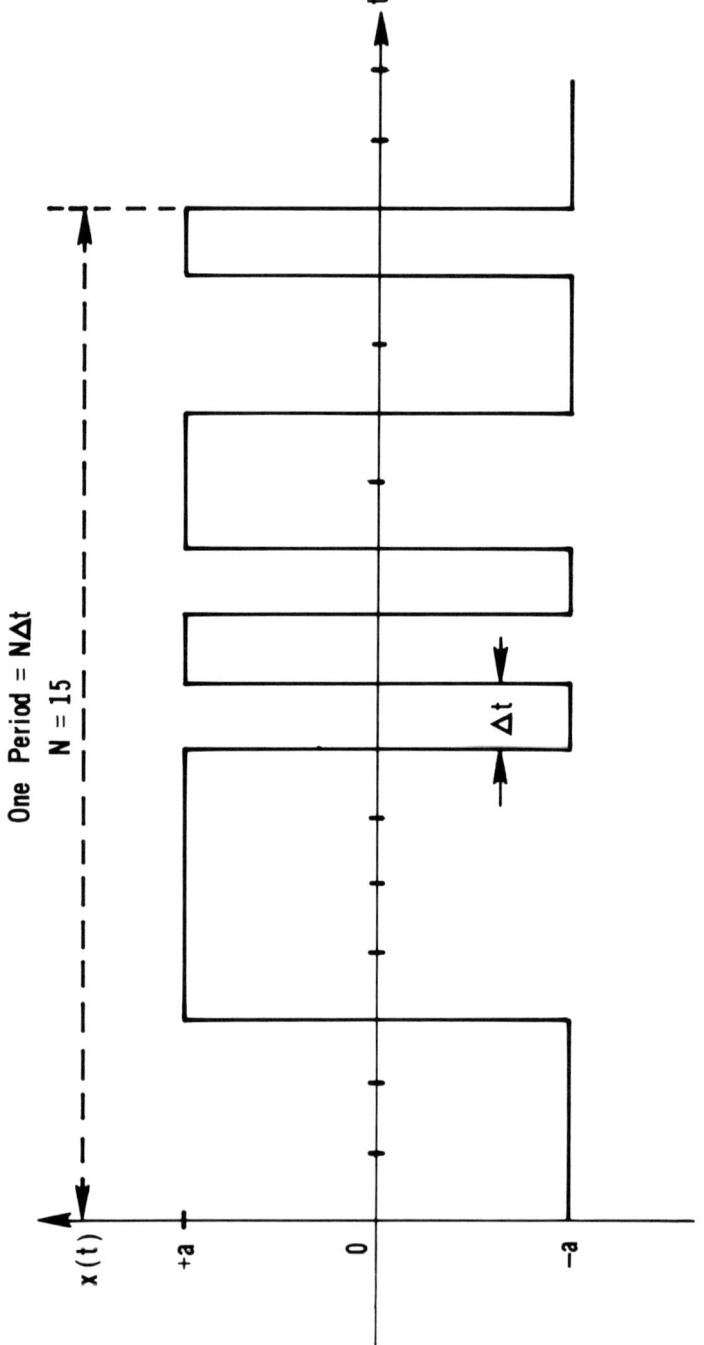

Figure 4-4. A 15-bit Pseudo-random Binary Sequence (PRBS)

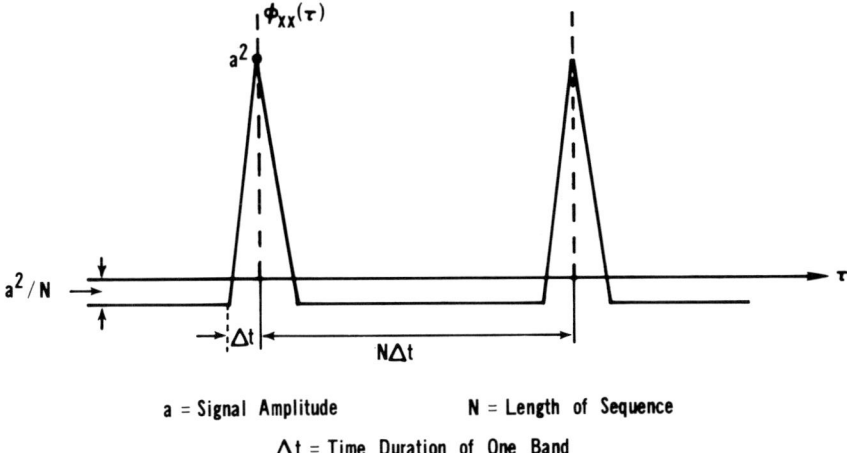

Figure 4–5. Autocorrelation Function of PRBS

Table 4–1
Sum-digit Table of a Module-two Adder

Input 1	Input 2	Module-two sum
0	0	0
0	1	1
1	0	1
1	1	0

function occurs that distorts the identification results. It is then obvious that systems having long settling times cannot be conveniently identified by the cross-correlation method using PRBS as a test signal.

For least-squares identification methods using PRBS as a test signal, there exists a unique computational advantage. Suppose the input $u(t)$ is PRBS, and that we sample the input-output in synchronism with Δt. In other words, the sampling instances coincide with the transition times of PRBS. Furthermore, the sampling period T of the weighting function $g(kT)$ is also chosen to be $T = \Delta t$. Because the sampled input autocorrelation function $\phi_{uu}(k)$ over one period is

$$\phi_{uu}(0) = a^2$$

and

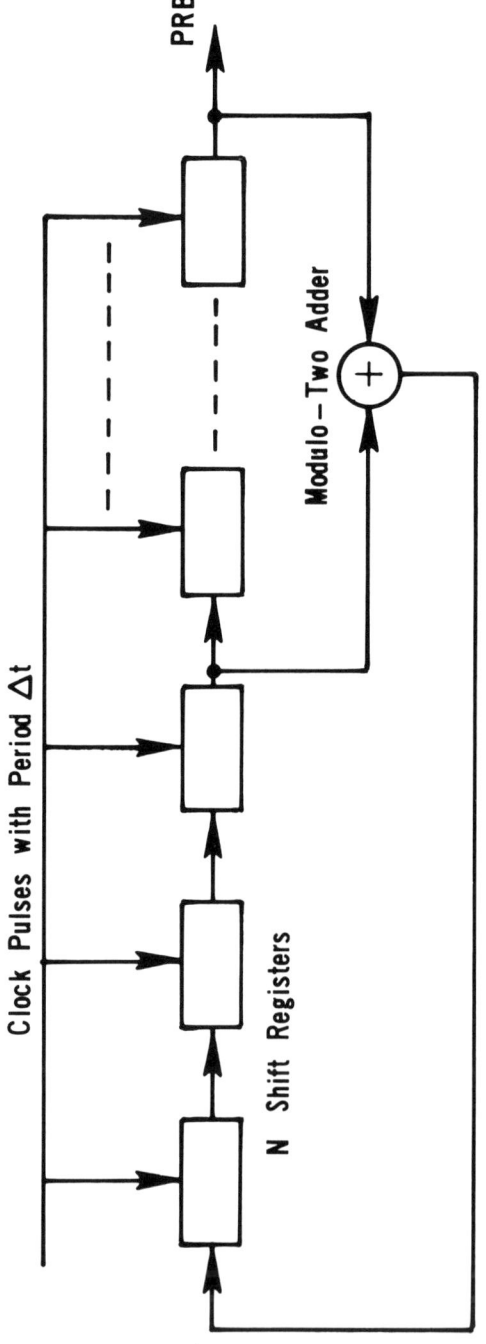

Figure 4-6. *N*-stage Shift Register for PRBS Generation

$$\phi_{uu}(k) = -\frac{a^2}{N}, \quad (k \neq 0)$$

where a is the amplitude of PRBS. The matrix Φ in equation (4.21) then has the known form

$$\Phi = a^2 \begin{bmatrix} 1 & -\frac{1}{N} & \cdots & -\frac{1}{N} \\ -\frac{1}{N} & 1 & & -\frac{1}{N} \\ \vdots & & & \vdots \\ -\frac{1}{N} & -\frac{1}{N} & \cdots & 1 \end{bmatrix}$$

It can be shown (Hill and McMurty 1964) that this Φ has an inverse as

$$\Phi^{-1} = \frac{N}{a^2(N+1)} \begin{bmatrix} 2 & 1 & \cdots & 1 \\ 1 & 2 & & 1 \\ \vdots & & & \vdots \\ 1 & & \cdots & 2 \end{bmatrix}$$

Thus the estimate $\hat{\mathbf{h}}$ becomes

$$\hat{\mathbf{h}} = \Phi^{-1}\Gamma = \frac{N}{a^2(N+1)} \begin{bmatrix} 2 & 1 & \cdots & 1 \\ 1 & 2 & & 1 \\ \vdots & & & \vdots \\ 1 & 1 & & 2 \end{bmatrix} \begin{bmatrix} \phi_{uy}(0) \\ \phi_{uy}(1) \\ \vdots \\ \phi_{uy}(p) \end{bmatrix} \quad (4.31)$$

This simplification is very significant because the costly inversion of Φ is eliminated. The corresponding improvement in computational efficiency can be considerable when the dimension p is large. Figure 4–7 illustrates a typical digital computer system for weighting function identification.

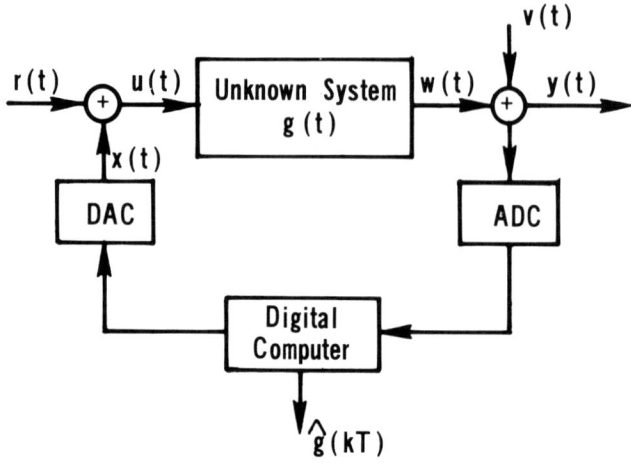

ADC, DAC Are Digital–Analog Converters
r(t) = Normal System Input
y(t) = Observable Output
x(t) = PRBS
v(t) = Disturbing Noise

Figure 4–7. A Digital Computer Identification Scheme

An Example

We will now present some experimental results of weighting function identification employing the computer system shown in figure 4–7 in conjunction with the PRBS testing signal (Hill and McMurty 1964). The experimental set-up is as follows:

1. The system input $r(t)$ is set to zero: $r(t) = 0$.
2. A second-order system is simulated by an analog computer with the transfer function as $H(s) = 2/(s^2 + s + 2)$.
3. The test signal $x(t)$ is a PRBS of 23 bits with magnitudes ± 1 and discrete interval length $\Delta t = 0.453$ seconds, which is sufficient to cover the significant portion of the weighting function for the system to be identified. The sampling rate for the output $y(t)$ is $T = \Delta t = 0.453$ seconds, which is kept in synchronism with the PRBS.
4. Twenty sequences of the above PRBN are used for each experimental run. Two cases for the output noise are considered:

a. $v(t) = 0$
b. $\sigma_v^2 = \sigma_y^2/1.9$ where σ_y^2 is the system output variance and $v(t)$ is a band-limited white noise.

The estimates $\hat{g}(iT)$, $i = 1, 2, \ldots p$ are computed from equation (4.31) letting $N = p = 23$, $a = 1$ and the relationship $\hat{g} = \hat{h}/T$. The identification results are shown in figure 4-8, where both the noise-free and noisy estimates are plotted along with the exact weighting function $g(t)$. It is seen that estimates $\hat{g}(t)$ are in good agreement with $g(t)$.

4.7 On-line Least-squares Identification

The basic least-squares estimation results presented in section 4.3 are of *one shot* or *batch processing* type in that the estimates are calculated based on the entire data record. This mode of solution is called *off-line*. If new data are measured and new parameter estimates are desired, the same solution procedure will have to be repeated with the new data added to the old record. Obviously, this mode of solution is not computationally desirable.

As we have shown in chapter 3, it is possible to derive a recursive algorithm for the least-squares solution of equation (4.11) so that the estimates $\hat{h}$ can be calculated sequentially. This means that we can perform on-line identification on **h**. The development of such an algorithm is briefly discussed below.

Suppose an estimate $\hat{\mathbf{h}}(m)$ has been obtained at the mth sampling instant based on

$$\hat{\mathbf{h}}(m) = (\mathbf{U}_m^T \mathbf{U}_m)^{-1} \mathbf{U}_m^T \mathbf{y}_m \tag{4.32}$$

in which the subscription on the right hand side indicates that the matrices U and y, as defined in equation (4.8), are formed by input-output data measured at the mth sampling instant and the past. Assume that a new input-output data pair $[u(m + 1), y(m + 1)]$ are measured at the $(m + 1)$th instant. Instead of recomputing equation (4.32) for $\hat{\mathbf{h}}(m + 1)$, we wish merely to update $\hat{\mathbf{h}}(m)$ based on the new data. Thus only simple calculations are needed to improve the parameter estimates at every sampling instant. This procedure is called *on-line identification*.

Recall that in chapter 3 we have developed a sequential least-squares estimation algorithm that is directly applicable to updating $\hat{\mathbf{h}}(m)$ in equation (4.32). Following the main results in equation (3.31), we can immediately write

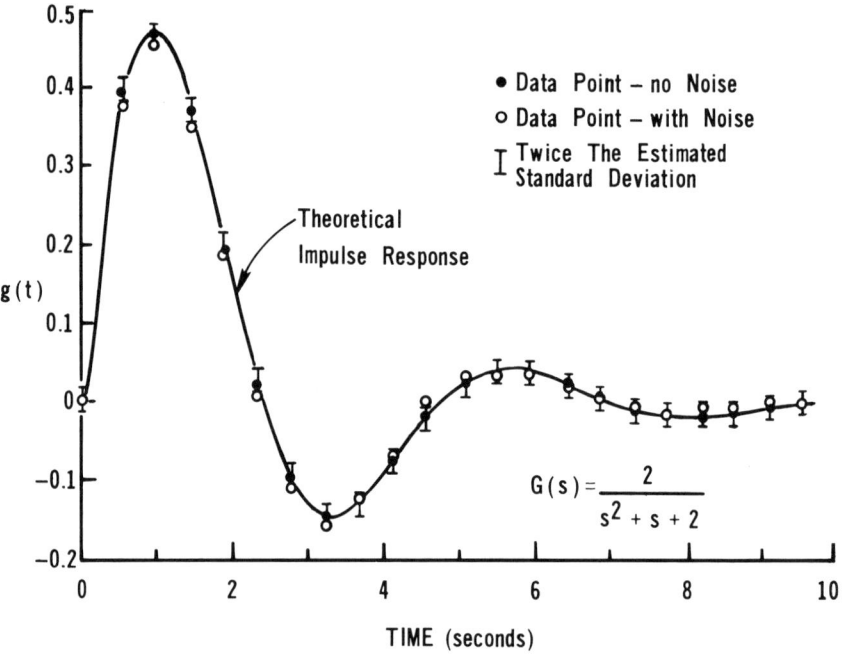

Figure 4-8. Identification Results of a Second-order Linear System

$$\hat{h}(m+1) = \hat{h}(m) + \gamma(m+1)P(m)x(m+1)$$
$$\cdot [y(m+1) - x^T(m+1)\hat{h}(m)]$$

$$P(m+1) = P(m) - \gamma(m+1)P(m)x(m+1)x^T(m+1)P(m) \quad (4.33)$$

$$\gamma(m+1) = 1/[1 + x^T(m+1)P(m)x(m+1)]$$

in which $x(m+1)$ is defined as

$$x(m+1) = [u(m+1), u(m), \ldots, u(m-p+1)]^T \quad (4.34)$$

The matrix $P(m)$ is defined by

$$P(m) = (U_m^T U_m)^{-1} \quad (4.35)$$

In view of equation (4.14), we see that $P(m)$ is a measure of the covariance Ψ of $\hat{h}(m)$ in that

$$\Psi = \sigma_v^2 (U_m^T U_m)^{-1} = \sigma_v^2 P(m) \quad (4.36)$$

By solving the pair $(\hat{h}, P)$ recursively we can update the system weighting sequence $\hat{h}$ at every sampling instant. Consequently, it is quite possible to continuously adjust $\hat{h}$ to track slowly time-varying systems.

We note that the on-line algorithms do not involve any matrix inversion, so they can be computed quickly. The initial values $\hat{\mathbf{h}}(0)$ and $\mathbf{P}(0)$ are either selected on the basis of some engineering guess, or obtained from a one-shot least-squares estimation (see chapter 3).

The on-line algorithm presented above has some interesting properties. For example, the algorithm can be interpreted as a form of gradient search operating in a learning model identification mode. To see this, we define an error signal $e(m + 1)$:

$$e(m + 1) = y(m + 1) - \mathbf{x}^T(m + 1)\hat{\mathbf{h}}(m) \qquad (4.37)$$

This error represents the difference between the measured output $y(m + 1)$ and the predicted output $\mathbf{x}^T(m + 1)\hat{\mathbf{h}}(m)$ from the system model wherein the latest parameter estimate $\hat{\mathbf{h}}(m)$ is used. Substituting this error into the on-line algorithm of equation (4.33), we get

$$\hat{\mathbf{h}}(m + 1) = \hat{\mathbf{h}}(m) + \mathbf{G}(m + 1)\mathbf{x}(m + 1)e(m + 1) \qquad (4.38)$$

where $\mathbf{G}(m + 1)$ is a gain matrix defined by

$$\mathbf{G}(m + 1) = \mathbf{P}(m)[1 + \mathbf{x}^T(m + 1)\mathbf{P}(m)\mathbf{x}(m + 1)]^{-1} \qquad (4.39)$$

We immediately recognize that equation (4.38) is a gradient algorithm minimizing the error criterion

$$e^2(m + 1) = [y(m + 1) - \mathbf{x}^T(m + 1)\hat{\mathbf{h}}(m)]^2$$

with respect to $\hat{\mathbf{h}}(m)$. This means that

$$\hat{\mathbf{h}}(m + 1) = \hat{\mathbf{h}}(m) - \frac{1}{2}\mathbf{G}(m + 1)\frac{\partial}{\partial \hat{\mathbf{h}}}e^2(m + 1),$$

$\mathbf{G}(m + 1)$ is a positive definite matrix, and

$$[1 + \mathbf{x}^T(m + 1)\mathbf{P}(m)\mathbf{x}(m + 1)]^{-1}$$

is a positive scalar.

The on-line identification algorithm can be conveniently depicted by the block diagram in figure 4–9. The error $e(m)$ is clearly seen to be the output difference between the system and the model. Therefore, $e(m)$ is viewed as an *output error* that is used to perturbed the model parameters. This arrangement of parameter adjustment is a called the *learning model identification*.

There are a variety of ways of choosing the adjustment gain $\mathbf{G}(m)$ in the gradient algorithm (Nagumo and Noda 1967; Mendel 1967). $\mathbf{G}(m)$ can be a scalar or a matrix; it can also be constant or time-varying. For the case in which the output noise $v(t)$ in the system is absent, a positive definite constant gain $\mathbf{G}(m) = \mathbf{G}$ is sufficient to guarantee the convergence of $\hat{\mathbf{h}}(m)$ to $\mathbf{h}$ as m tends to infinity. However, the convergence fails when noise $v(t)$ is present. In this case, only a variable gain of special properties

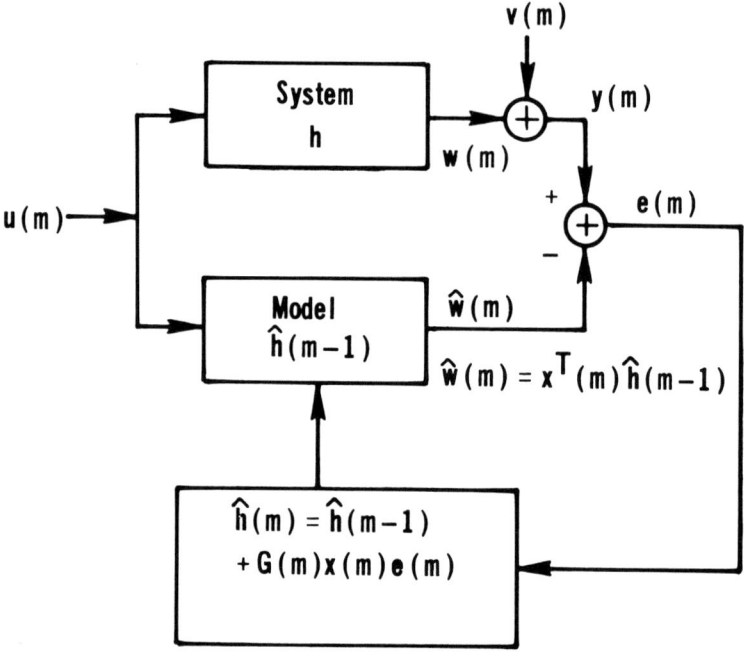

Figure 4-9. Learning Model Identification Interpretation of the On-line Least-squares Estimation Algorithm

can lead to convergence. These properties are specified in the theory of stochastic approximation (Graupe 1972; Eykhoff 1974). A monotonically decreasing gain at the rate of $1/m$ satisfies the requirements of stochastic approximation. Since the matrix $\mathbf{P}(m)$ in equation (4.39) is decreasing in the order of $1/m$ for large m [see equations (4.36) and (4.16)], the least-squares algorithm in equation (4.38) can be viewed as a form of stochastic approximation.

4.8 Multivariable System Identification

The identification techniques presented so far are based on single-input single-output systems. However, the results can be easily extended to systems with multiple inputs and multiple outputs. As mentioned in chapter 2, a linear continuous multivariable system can be represented by a weighting function matrix

$$\mathbf{G}(t) = [g_{ij}(t)] \quad i = 1, 2, \ldots, q \quad (4.40)$$
$$j = 1, 2, \ldots, r$$

This system has q outputs and r inputs as shown in figure 4–10. In this system, each output is connected to all the inputs through the weighting function $g_{ij}(t)$. Thus

$$w_i(t) = \sum_{j=1}^{r} \int_0^t g_{ij}(t - \tau) u_j(\tau) \, d\tau \quad i = 1, 2, \ldots, q \quad (4.41)$$

and we are required to estimate a total of qxr weighting functions $g_{ij}(t)$.

To apply the single-variable techniques to this multivariable problem, we likewise approximate each weighting function $g_{ij}(t)$ by a finite sequence $\{g_{ij}(kT)\}$, $k = 1, 2, \ldots, p$. Define the corresponding weighting sequence $\{h_{ij}(kT)\}$ by $h_{ij}(kT) = Tg_{ij}(kT)$. Then we have

$$y_i(kT) = \sum_{j=1}^{r} \sum_{\lambda=k-p}^{k} h_{ij}(kT - \lambda T) u_j(\lambda T) + v_i(kT) \quad (4.42)$$

From this point on we can follow the single-variable formulation to identify the sequences $\{h_{ij}(k)\}$. Define the following:

$$\mathbf{h}_{ij} = [h_{ij}(0), h_{ij}(1), \ldots, h_{ij}(p)]^T$$
$$\mathbf{h}_i = [\mathbf{h}_{i1}^T, \mathbf{h}_{i2}^T, \ldots, \mathbf{h}_{ir}^T]^T$$

$$\mathbf{U}_j = \begin{bmatrix} u_j(p) & u_j(p-1) & \cdots & u_j(0) \\ u_j(p+1) & u_j(p) & \cdots & u_j(1) \\ \vdots & & & \\ u_j(p+m) & u_j(p+m-1) & & u_j(m) \end{bmatrix}$$

$$\mathbf{y}_i = [y_i(p), y_i(p+1), \ldots, y_i(p+m)]^T$$
$$\mathbf{v}_i = [v_i(p), v_i(p+1), \ldots, v_i(p+m)]^T$$

We can set up a vector equation for each output based on $m + 1$ set of observations from equation (4.42):

$$\mathbf{y}_i = [\mathbf{U}_1, \mathbf{U}_2, \ldots, \mathbf{U}_r] \mathbf{h}_i + \mathbf{v}_i$$

or

$$\mathbf{y}_i = \mathbf{U}\mathbf{h}_i + \mathbf{v}_i$$

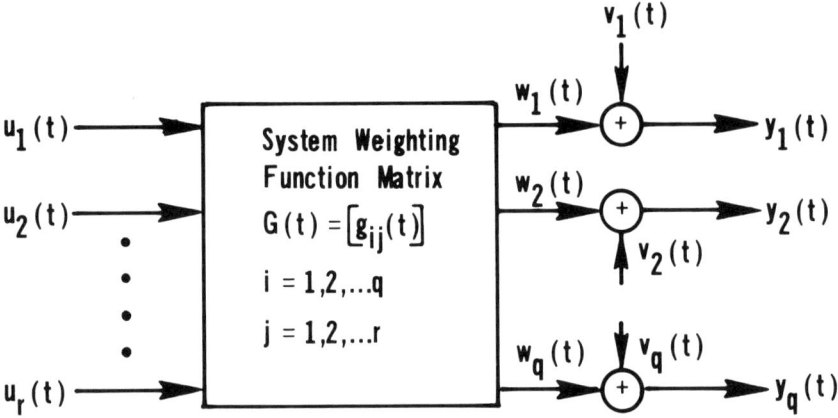

Figure 4-10. Multi-input and Multi-output Linear System; $g_{ij}(t)$ is the Weighting Function between Input $u_j(t)$ and Output $w_i(t)$

Now the vector $\mathbf{h}_i$ can be estimated by

$$\hat{\mathbf{h}}_i = (\mathbf{U}^T\mathbf{U})^{-1}\mathbf{U}^T\mathbf{y}_i \tag{4.43}$$

We observe that $\hat{\mathbf{h}}_i$ contains estimates of r weighting functions. Therefore, by repeating the above solution q times (that is, let $i = 1, 2, \ldots, q$), we can complete the identification of the entire weight function matrix $\mathbf{G}(t)$.

It is apparent that the computations involved in equation (4.43) can be excessive when r and q are large. This difficulty can be minimized, however, by letting the inputs $u_j(t)$ be stationary random processes that are uncorrelated with one another, that is $E[u_i u_j] = 0$ for $i \neq j$. In this case $\mathbf{U}^T\mathbf{U}$ is simplified to the following diagonal form when the data length m is large:

$$\mathbf{U}^T\mathbf{U} = \begin{bmatrix} \mathbf{U}_1^T\mathbf{U}_1 & 0 & \cdots & 0 \\ 0 & \mathbf{U}_2^T\mathbf{U}_2 & & \vdots \\ \vdots & & \ddots & \vdots \\ 0 & & & \mathbf{U}_r^T\mathbf{U}_r \end{bmatrix}$$

Thus equation (4.43) can be decomposed into

$$\hat{\mathbf{h}}_{ij} = (\mathbf{U}_j^T\mathbf{U}_j)^{-1}\mathbf{U}_j^T\mathbf{y}_i \tag{4.44}$$

where $i = 1, 2, \ldots, q$; and $j = 1, 2, \ldots, r$. This result is analogous to equation (4.11) for single-variable systems. So we see that the dimensionality in equation (4.44) is drastically reduced. However, it requires $q \times r$ separate computations to complete the job.

As shown in section 4.4, the computation of equation (4.44) can be further simplified if the inputs are white noises. Following the result in equation (4.23), the elements in $\hat{h}_{ij}$ can be expressed as

$$\hat{h}_{ij}(\lambda) = \frac{1}{G_j} \phi_{ij}(\lambda) \quad \lambda = 0, 1, 2, \ldots, p \tag{4.45}$$

in which G_j is the mean square value of $u_j(k)$

$$G_j = \frac{1}{m+1} \sum_{k=0}^{m} u_j^2(k)$$

and $\phi_{ij}(\lambda)$ is the empirical cross-correlation function between $u_j(k)$ and $y_i(k)$:

$$\phi_{ij}(\lambda) = \frac{1}{m+1} \sum_{k=\tau}^{m+\tau} y_i(k) u_j(k - \lambda)$$

where τ is a suitable positive integer such that $\tau \geq \lambda$. The required random inputs can be realized by using independently generated PRBSes.

In summary, we see that multivariable systems identification is no more difficult than single-variable systems identification except that a matrix of weighting functions are to be estimated. Equation (4.43) provides the estimates of r weighting functions that make up one row in the weighting function matrix. However, it is possible to estimate just one weighting function at a time, as indicated by equation (4.44), if the inputs are statistically independent from one another. Finally we have shown in equation (4.45) that each sample point on a weighting function can be individually computed if each input is a white random process.

4.9 Concluding Remarks

In this chapter, we have defined the basic problem of weighting function identification and given the solution by means of least-squares parameter estimation. We consider the identification completed when sufficient sample points on the weighting function are computed.

It has been demonstrated that the least-squares procedure is applicable to both single-variable and multivariable linear systems. We have shown that identification procedures can be carried out either off-line or on-line. Moreover, relationships of these algorithms to those of cross-correlation and iterative gradient search are established.

The problem of optimum test signal has been discussed in detail, and it was shown that white noise is the optimum among signals having the same average power. We have indicated that PRBS is a practical substitute for white noise in system identification.

Finally, we wish to mention that weighting functions can be approximated by finite orthogonal series. Identification of such approximate models has been proven very effective. In particular, least-squares algorithms introduced in this chapter can be readily applied to estimating the coefficients in the orthogonal series expansion from sampled input-output data (Levin 1960). An important feature of such a model is that it provides a continuous representation of an unknown weighting function rather than a finite sampled data representation.

References

Anderson, G. W.; Buland, R. N.; and Cooper, G. R., "Use of Cross-Correlation in an Adaptive Control System," *Proceedings of the National Electronics Conference*, Vol. 15, Oct. 1959.

Davies, W. D. T., *System Identification for Self-adaptive Control*, Wiley-Interscience, New York, 1970.

Eykhoff, P., *System Identification*, Wiley, New York, 1974.

Goodwin, G. C.; Murdock, J. C.; and Payne, R. L., "Optimal Test Signal Design for Linear Single-input Single-output System Identification," *International Journal of Control*, Vol. 17, No. 1, pp. 45–55, 1973.

Graupe, D., *Identification of Systems*, Van Nostrand Reinhold, New York, 1972.

Hill, J. D., and McMurtry, G. J., "An Application of Digital Computers to Linear System Identification," *IEEE Transactions on Automatic Control*, Vol. AC-9, pp. 536–538, Oct. 1964.

Levin, M. J., "Optimum Estimation of Impulse Response in the Presence of Noise," *IRE Transactions on Circuit Theory*, Vol. CT-7, pp. 50–56, Mar. 1960.

Mehra, R. K., "Optimal Input Signals for Parameter Estimation in Dynamic Systems—Survey and New Results," *IEEE Transactions on Automatic Control*, Vol. AC-19, pp. 753–768, Dec. 1974.

Mendel, J. M., "Gradient, Error-correction Identification Algorithms," *Information Science*, Vol. 1, pp. 23–42, 1968.

Nagumo J., and Noda, A., "A Learning Method of System Identification," *IEEE Transactions on Automatic Control*, Vol. 12, AC-12, pp. 282–287, 1967.

Turin, G. L., "On the Estimation in the Presence of Noise of the Impulse Responce of a Random Linear Filter," *IRE Transactions on Information Theory*, Vol. II-3, No. 1, Mar. 1957.

Wiener, N., *The Extrapolation, Interpolation, and Smoothing of Stationary Time Series with Engineering Applications*, Wiley, New York, 1949.

Appendix 4A
PRBS Generation
Computer Program

```
C  U PRBS, LENGTH 127, AMPLITUDE 0.5
C  NDP NUMBER OF DATA POINTS
   DIMENSION U(NDP), SH(7)
   DO 2 I=1,7
   SH(I) = 1.0
 2 CONTINUE
   DO 4 I=1, NDP
   S=SH(7)+SH(6)
   IF (S.GT.1.5) S=0.0
   DO 3 J=2,7
   JM=8-J
   SH(JM+1)=SH(JM)
 3 CONTINUE
   SH(1)=S
   U(I)=S-0.5
 4 CONTINUE
```

5
Linear Parametric Model Identification by Least Squares

5.1 Introduction

This chapter examines linear discrete systems modeled by difference equations, transfer functions, and state variable equations. All of these models are called parametric models. In contrast, the weighting function model discussed in the last chapter is nonparametric.

Parametric system models are usually more useful because modern control theory and system design techniques require the state variable description of the system dynamics. Although systems we encounter are generally continuous, it is more practical and convenient to approximate these systems by discrete models. These models are appropriate if only the relationships between the sampled input and sampled output signals are of concern. We understand that whenever digital computer techniques are employed for control or data storage and processing, it is necessary to deal with all the signals in their discrete forms. Another reason that we prefer discrete models over continuous models is that difference equations are algebraic in nature and hence are easier to manipulate and identify than the differential equations.

Nevertheless, continuous systems identification is treated in this chapter as an extension of techniques developed for discrete systems. We shall see that differential equations can be formulated in a form similar to difference equations for the purpose of identification if additional filtering procedures are incorporated.

Based on our discussions in chapter 2, it is obviously possible to identify the weighting sequence model first, and then fit it by a difference equation using their mathematical equivalence relations. However, this indirect procedure of identifying a parametric model is cumbersome and limited to only single-variable systems. Therefore, a direct identification approach is desirable. The suggested indirect approach will be explored fully in chapter 7 as an alternative means to deal with noisy systems.

The identification problem discussed in this chapter is a simple one in that we consider that the system is nearly noise-free. Therefore, it is possible to assume white random errors in the difference equation model. This situation then allows the direct application of the simple least-squares technique to estimate the model parameters. The noisy system case will be considered in chapter 6.

5.2 The Basic Identification Problem

We begin by considering a single-variable linear time-invariant discrete system as shown in figure 5–1. $u(kT)$ and $y(kT)$ are the samples of the system's input and output signals where T is the constant sampling period. Let the system be described by an nth order difference equation of constant coefficients:

$$y(k) + a_1 y(k-1) + a_2 y(k-2) + \ldots + a_n y(k-n)$$
$$= b_0 u(k) + b_1 u(k-1) + \ldots + b_n u(k-n). \quad (5.1)$$

In the above equation, T is dropped from $u(kT)$ and $y(kT)$ for the sake of simplicity. As suggested in chapter 2, this system can be rewritten in the following shifting-operator form:

$$A(q^{-1})y(k) = B(q^{-1})u(k) \quad (5.2)$$

where

$$A(q^{-1}) = 1 + a_1 q^{-1} + a_2 q^{-2} + \ldots + a_n q^{-n}$$
$$B(q^{-1}) = b_0 + b_1 q^{-1} + b_2 q^{-2} + \ldots + b_n q^{-n}$$

As shown in chapter 2, this system has a transfer function $H(z)$ defined by

$$H(z) = \frac{B(z^{-1})}{A(z^{-1})} \quad (5.3)$$

With the system equation (5.1) in mind, we now state the *basic system identification problem* as follows.

Assume that the system in equation (5.1) is stable and the order is n, and the system is in steady state operation. Given n and the input-output measurements $\{u(k), y(k)\}$, $k = 1, \ldots, N + n$, estimate the constant system parameters a_i and b_i.

We will solve the above problem by fitting the system equation to the input-output data so as to choose the best parameter values in the sense of minimum-square-error. To accomplish this, we first write the system equation as

$$A(q^{-1})y(k) = B(q^{-1})u(k) + e(k) \quad (5.4)$$

in which $y(k)$, $u(k)$ are the measured data. The term $e(k)$ is introduced to account for the *fitting error*. In the theory of linear regression, $e(k)$ is referred to as the *residual*. In terms of system identification error classification as discussed in chapter 1, $e(k)$ is called the *equation error*. This error is schematically depicted in figure 5–2. We develop the parameter estimation procedures in the following section.

Figure 5–1. A Stable Single-variable Discrete System

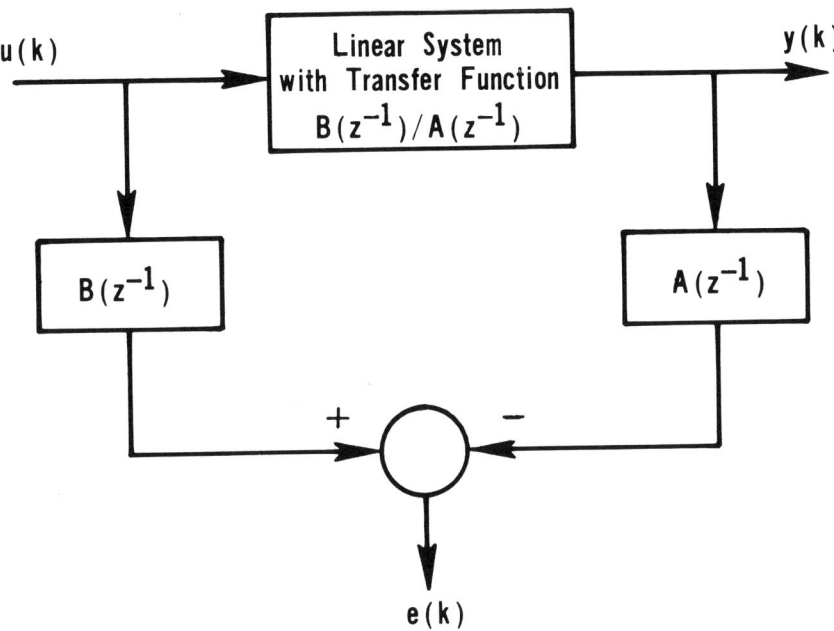

Figure 5–2. Block Diagram Representation of the Error Signal $e(k)$. The Blocks Containing $A(z^{-1})$ and $B(z^{-1})$ Can Be Regarded as Filters

5.3 Least-squares Solution

Rewrite equation (5.4) in the following form:

$$y(k) = -\sum_{i=1}^{n} a_i y(k - i) + \sum_{i=0}^{n} b_i u(k - i) + e(k) \qquad (5.5)$$

Define the $2n + 1$ input-output vector $x(k)$ as

$$\mathbf{x}(k) = [-y(k - 1), \ldots, -y(k - n), u(k), \ldots, u(k - n)]^T$$

and the n parameter vector $\boldsymbol{\theta}$ as

$$\boldsymbol{\theta} = [a_1, \ldots, a_n, b_0, \ldots, b_n]^T$$

Then we can write
$$y(k) = \mathbf{x}^T(k)\boldsymbol{\theta} + e(k) \tag{5.6}$$

Because we have available a string of data $\{y(k), u(k)\}$ for $k = 1, \ldots, (N + n)$, we can set up a system of N equations as (assume $N \gg 2n$)
$$\mathbf{y} = \mathbf{X}\boldsymbol{\theta} + \mathbf{e} \tag{5.7}$$

where
$$\mathbf{y} = [y(n + 1), y(n + 2), \ldots, y(n + N)]^T$$
$$\mathbf{e} = [e(n + 1), e(n + 2), \ldots, e(n + N)]^T$$

$$\mathbf{X} = \begin{bmatrix} \mathbf{x}^T(n+1) \\ \mathbf{x}^T(n+2) \\ \vdots \\ \mathbf{x}^T(n+N) \end{bmatrix} = \begin{bmatrix} -y(n), \ldots, & -y(1), & u(n+1), \ldots, u(1) \\ -y(n+1), \ldots, & -y(2), & u(n+2), \ldots, u(2) \\ \vdots & & \vdots \\ -y(n+N-1), \ldots, & -y(N), & u(n+N), \ldots, u(N) \end{bmatrix}$$

With the vector equation (5.7), in which $\mathbf{y}$ and $\mathbf{X}$ are given, established, we can conveniently estimate the parameter vector $\boldsymbol{\theta}$ by means of least squares. This identification approach was first suggested by Kalman (1958). The complete solution is presented below.

The least-squares procedure says that we choose the estimate of $\boldsymbol{\theta}$ as that value of $\boldsymbol{\theta}$ that minimizes the error function J:

$$J = \sum_{k=n+1}^{N+n} e^2(k) = \mathbf{e}^T\mathbf{e}$$
$$= (\mathbf{y} - \mathbf{X}\boldsymbol{\theta})^T(\mathbf{y} - \mathbf{X}\boldsymbol{\theta}) \tag{5.8}$$

Upon setting
$$\left.\frac{\partial J}{\partial \boldsymbol{\theta}}\right|_{\boldsymbol{\theta}=\hat{\boldsymbol{\theta}}} = 0$$

we immediately obtain the least-squares estimate $\hat{\boldsymbol{\theta}}$ by
$$\hat{\boldsymbol{\theta}} = (\mathbf{X}^T\mathbf{X})^{-1}\mathbf{X}^T\mathbf{y} \tag{5.9}$$

This solution exists if $\mathbf{X}^T\mathbf{X}$ is nonsingular. We assume that this condition is met by the input-output data (we have already assumed that the system is stable with exact order n, and the data length $N \gg 2n$). The type of input sequence $\{u(k)\}$ driving the system that yields a nonsingular $(2n + 1) \times (2n + 1)$ matrix $\mathbf{X}^T\mathbf{X}$ is called *persistently exciting* of order $2n + 1$. The detailed structures of the matrices $\mathbf{X}^T\mathbf{X}$ and $\mathbf{X}^T\mathbf{y}$ are shown in equations (5.10) and (5.11).

$$X^TX = \left[\begin{array}{cccc|cccc}
\sum\limits_{k=n}^{N+n-1} y^2(k) & \sum\limits_{k=n}^{N+n-1} y(k)y(k-1) & \cdots & \sum\limits_{k=n}^{N+n-1} y(k)y(k-n+1) & -\sum\limits_{k=n}^{N+n-1} y(k)u(k+1) & \cdots & -\sum\limits_{k=n}^{N+n-1} y(k)u(k) & -\sum\limits_{k=n}^{N+n-1} y(k)u(k-n+1) \\
 & \sum\limits_{k=n-1}^{N+n-2} y^2(k) & \cdots & \sum\limits_{k=n-1}^{N+n-2} y(k)y(k-n+2) & -\sum\limits_{k=n-1}^{N+n-2} y(k)u(k+2) & \cdots & & -\sum\limits_{k=n-1}^{N+n-2} y(k)u(k-n+2) \\
 & & \ddots & \vdots & \vdots & & & \vdots \\
 & & & \sum\limits_{k=1}^{N} y^2(k) & -\sum\limits_{k=1}^{N} y(k)u(k+n-1) & \cdots & -\sum\limits_{k=1}^{N} y(k)u(k+n) & -\sum\limits_{k=1}^{N} y(k)u(k) \\
\hline
 & & & & \sum\limits_{k=n+1}^{N+n} u^2(k) & \sum\limits_{k=n+1}^{N+n} u(k)u(k-1) & \cdots & \sum\limits_{k=n+1}^{N+n} u(k)u(k-n+1) \\
 & & & & & \sum\limits_{k=n}^{N+n-1} u^2(k) & \cdots & \sum\limits_{k=n}^{N+n-1} u(k)u(k-n+2) \\
 & & & & & & \ddots & \vdots \\
 & & & & & & & \sum\limits_{k=1}^{N} u^2(k)
\end{array}\right]$$

(5.10)

$$\mathbf{X}^T\mathbf{y} = \begin{bmatrix} -\sum_{k=n+1}^{N+n} y(k)y(k-1) \\ -\sum_{k=n+1}^{N+n} y(k)y(k-2) \\ \vdots \\ -\sum_{k=n+1}^{N+n} y(k)y(k-n) \\ \hline \sum_{k=n+1}^{N+n} y(k)u(k) \\ \sum_{k=n+1}^{N+n} y(k)u(k-1) \\ \vdots \\ \sum_{k=n+1}^{N+n} y(k)u(k-n) \end{bmatrix} \qquad (5.11)$$

5.4 Statistical Properties of Parameter Estimates

In this section, we examine the existence and accuracy of the least-squares parameter estimates. A noise interpretation for the fitting error is also given.

First, we see that for $\hat{\boldsymbol{\theta}}$ to exist, it is necessary that the matrix $\mathbf{X}^T\mathbf{X}$ has an inverse for large data length N. For this condition to be met, it is required that the input sequence $\{u(k)\}$ be persistently exciting of order $2n + 1$. As before, PRBS serves as an optimum test signal.

Second, in order to examine the statistical accuracy of $\hat{\boldsymbol{\theta}}$, we need to know the statistics of the fitting error, or residual, $e(k)$. Let us assume that $e(k)$ is an independent random variable with zero mean and variance σ^2, that is

$$E[e(k)] = 0 \qquad E[e(i)e(j)] = \sigma^2 \delta_{ij}.$$

Furthermore, we assume that $e(k)$ is uncorrelated with $u(k)$ and $y(k)$.

First we wish to examine the biasness of $\hat{\boldsymbol{\theta}}$. We can express the estimator bias as

$$E[\hat{\boldsymbol{\theta}} - \boldsymbol{\theta}] = E[(\mathbf{X}^T\mathbf{X})^{-1}\mathbf{X}^T\mathbf{e}].$$

Since $\mathbf{X}$ and $\mathbf{e}$ are uncorrelated,

$$E[\hat{\boldsymbol{\theta}} - \boldsymbol{\theta}] = E[(\mathbf{X}^T\mathbf{X})^{-1}\mathbf{X}^T]E[\mathbf{e}] = 0.$$

Therefore $\hat{\boldsymbol{\theta}}$ is unbiased. That is, $E[\hat{\boldsymbol{\theta}}] = \boldsymbol{\theta}$.

Next, let us evaluate the covariance of $\hat{\boldsymbol{\theta}}$. We can show that (see chapter 3)

$$\boldsymbol{\Psi} = E[(\hat{\boldsymbol{\theta}} - \boldsymbol{\theta})(\hat{\boldsymbol{\theta}} - \boldsymbol{\theta})^T] = \sigma^2(\mathbf{X}^T\mathbf{X})^{-1}.$$

The diagonal elements of $\boldsymbol{\Psi}$ represent the variances of $\hat{\boldsymbol{\theta}}$. Also, an unbiased estimate of σ^2 can be obtained by

$$\hat{\sigma}^2 = \frac{1}{N}[\mathbf{y}^T\mathbf{y} - \mathbf{y}^T\mathbf{X}\hat{\boldsymbol{\theta}}].$$

Since we have assumed that the input is persistently exciting, the limit

$$\lim_{N \to \infty} \frac{1}{N}\mathbf{X}^T\mathbf{X} = \boldsymbol{\Phi}$$

exists and $\boldsymbol{\Phi}$ is positive definite. Then

$$\lim_{N \to \infty} \boldsymbol{\Psi} = \lim_{N \to \infty} \frac{\sigma^2}{N} \lim_{N \to \infty} (\mathbf{X}^T\mathbf{X})^{-1} = 0$$

which implies that $\hat{\boldsymbol{\theta}}$ converges to $\boldsymbol{\theta}$ as $N \to \infty$. Hence, $\hat{\boldsymbol{\theta}}$ is a consistent estimator of $\boldsymbol{\theta}$. The system is identifiable.

As pointed out in chapter 3, $\hat{\boldsymbol{\theta}}$ is equivalent to the maximum likelihood estimator if the residuals $e(k)$ are normally distributed random variables. It then can be stated that $\hat{\boldsymbol{\theta}}$ is asymptotically normal with mean $\boldsymbol{\theta}$ and covariance $\boldsymbol{\Psi}$.

Last, we wish to interpret the random fitting error $e(k)$ in terms of an output measurement noise. Shown in figure 5-3 is a system with noise $v(k)$ at the output. We wish to relate $v(k)$ to $e(k)$. From figure 5-3, we have

$$A(q^{-1})w(k) = B(q^{-1})u(k) \tag{5.12}$$

$$y(k) = w(k) + v(k)$$

Simplifying yields

$$A(q^{-1})y(k) = B(q^{-1})u(k) + A(q^{-1})v(k) \tag{5.13}$$

Comparing equations (5.13) and (5.4), we immediately see that

$$e(k) = A(q^{-1})v(k) \tag{5.14}$$

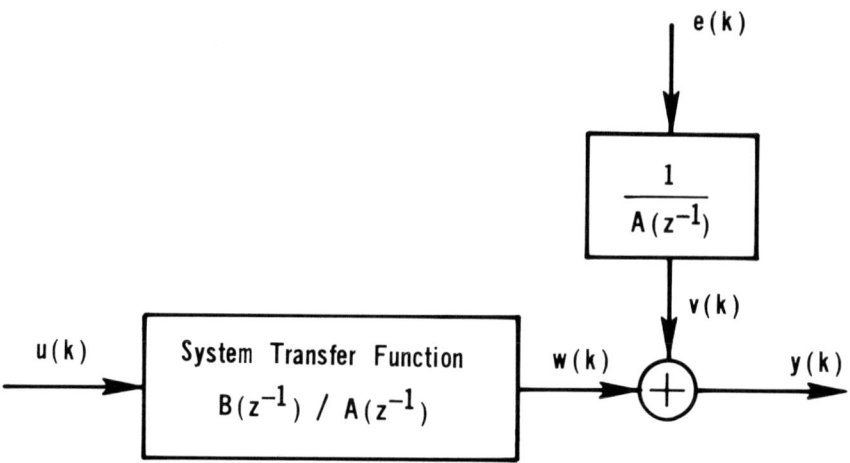

Figure 5-3. Systems with Output Noise of Special Type

which means that $v(k)$ is the output of a filter $1/A(z^{-1})$ with $e(k)$ as input. The significance of this analysis is that the identification of the noisy system in figure 5-3 is no different from that of the system in figure 5-2 so long as $v(k)$ is a filtered white noise with a special filter transfer function $1/A(z^{-1})$. The case of $v(k)$ with a general filter is discussed in detail in chapter 6.

An Example

To illustrate the performance of the least-square method, we consider the second-order system

$$y(k) - 1.5y(k-1) + 0.7y(k-2) = 1.0u(k-1) + 0.5u(k-2) + e(k)$$

(5.15)

This system is simulated on a digital computer to generate the output data $y(k)$ by the recursive formula

$$y(k) = 1.5y(k-1) - 0.7y(k-2) + 1.0u(k-1) + 0.5u(k-2) + e(k)$$

The input sequence $\{u(k)\}$ is a PRBN of amplitude 1, and the noise $\{e(k)\}$ is a sequence of normal $(0, \sigma)$ random numbers with σ adjustable.

We notice that $b_0 = 0$ in the system, so we do not include b_0 in the parameter set. Thus

$$\theta^T = [a_1, a_2, b_1, b_2] = [-1.5, 0.7, 1.0, 0.5].$$

Table 5-1 lists the least-squares estimates $\hat{\theta}$ for various levels of noise σ.

Table 5-1
Least-squares Parameter Estimates $\hat{\theta}$ Obtained for Data Length $N = 100$ at Various Noise Levels σ

Noise levels	Parameter estimates			
	$\hat{a}_1$	$\hat{a}_2$	$\hat{b}_1$	$\hat{b}_2$
$\sigma = 0.0$	-1.50 ± 0.00	0.70 ± 0.00	1.00 ± 0.00	0.50 ± 0.00
$\sigma = 0.1$	-1.50 ± 0.01	0.69 ± 0.01	0.99 ± 0.01	0.49 ± 0.02
$\sigma = 0.5$	-1.48 ± 0.04	0.67 ± 0.03	0.96 ± 0.06	0.48 ± 0.07
$\sigma = 1.0$	-1.47 ± 0.06	0.66 ± 0.06	0.93 ± 0.12	0.46 ± 0.14
$\sigma = 5.0$	-1.48 ± 0.07	0.74 ± 0.08	0.98 ± 0.61	0.41 ± 0.61
True parameter values	-1.50	0.70	1.00	0.50

Source: Adapted from data in K.J. Åström and P. Eykhoff, "System Identification—A Survey," *Automatica*, vol. 7, pp. 123–162, 1971.

The data length is $N = 100$ for all cases. The results show that exact parameter values can be obtained when noise is absent. As expected, the variance of $\hat{\theta}$ increases as noise level σ increases.

5.5 On-line Least-squares Identification

The estimation of θ can be carried out on-line when the solution in equation (5.9) is converted to a sequential form in terms of data length N. The derivation of the on-line algorithm is identical to that given in chapter 3. In view of equation (3.31), we can write $\hat{\theta}$ recursively as

$$\hat{\theta}(N + 1) = \hat{\theta}(N) + \gamma(N + 1)\mathbf{P}(N)\mathbf{x}(N + 1)$$
$$\cdot [y(N + 1) - \mathbf{x}^T(N + 1)\hat{\theta}(N)]$$
$$\mathbf{P}(N + 1) = \mathbf{P}(N) - \gamma(N + 1)\mathbf{P}(N)\mathbf{x}(N + 1)\mathbf{x}^T(N + 1)\mathbf{P}(N) \quad (5.16)$$
$$\gamma(N + 1) = 1/[1 + \mathbf{x}^T(N + 1)\mathbf{P}(N)\mathbf{x}(N + 1)]$$

where
$$\mathbf{x}(N + 1) = [-y(N), \ldots,$$
$$-y(N - n + 1), u(N + 1), \ldots, u(N - n + 1)]^T$$

As shown earlier, $\mathbf{P}(N)$ is a measure of the covariance of $\hat{\theta}(N)$. The choice of initial values $\hat{\theta}(0)$ and $\mathbf{P}(0)$ for starting up the algorithm can be made in the same ways as suggested in section 3.4 of chapter 3. The on-line algorithm allows us to easily update the estimate $\hat{\theta}$ as the number of measurements N increases. Thus it is possible to track slowly-varying system parameters.

5.6 System Order Determination

The system identification method discussed so far assumed that the order of the dynamic model is known a priori so that only the coefficients in the difference equation are to be estimated. In practice, however, the order of the system is seldom exactly known. Therefore, order determination is an important element of the system identification problem. Model order determination is also referred to as *system structure determination* in that the knowledge of the order n is directly related to the exact structure of a linear difference equation. Experiences have shown that a model of incorrect order can cause serious problems in control system design. Hence, it is important to make tests whenever necessary during system identification on the adequacy of the model order. We present two simple testing techniques for order determination. In addition, we discuss a recursive algorithm for computing least-squares estimates with increasing order.

A survey of literature (Unbehanen and Göbring 1974) shows that a variety of approaches are available for testing the adequacy of the system order n as described in equation (5.5). However, we will only discuss two practical techniques to illustrate how the order determination problem can be dealt with.

Testing the Goodness of Fit of the Model

A simple but very effective method for order testing is to compare the goodness of fit of the model to observed data for different model orders n. The goodness of fit is measured by the error-square-sum function J

$$J = (\mathbf{y} - \mathbf{X}\hat{\boldsymbol{\theta}})^T(\mathbf{y} - \mathbf{X}\hat{\boldsymbol{\theta}}) \tag{5.17}$$

where $\hat{\boldsymbol{\theta}}$ is the least-squares parameter estimate for a given model order n. In general, J decreases as n increases. However, the reduction of J ceases to be significant when n becomes greater than the true order n_0. This principle can be conveniently used to determine the desirable order of the model. Thus the procedure for order determination is simply to compute the least-squares parameter estimates $\hat{\boldsymbol{\theta}}$ and the corresponding error function J for a sequence of model orders $n = 1, 2, 3, \ldots$. The appropriate model order can be chosen as the one that J ceases to decrease significantly.

To illustrate how this testing procedure actually works, we consider again the second-order example system in equation (5.15). Suppose we fit the simulated input-output data by the model

$$y(k) = -\sum_{i=1}^{n} a_i y(k - i) + \sum_{i=1}^{n} b_i u(k - i) + e(k)$$

and let n take on values $n = 1, 2, 3$. Table 5–2 shows the values of J for each n and five levels of noise σ. We can see clearly that the correct order for all cases is $n = 2$.

We wish to note here that for the noise-free case, that is $\sigma = 0$, J becomes zero at $n = 2$, as it should. However, the estimate $\hat{\theta}$ does not exist for $n > 2$ due to the singularity of the matrix $\mathbf{X}^T\mathbf{X}$ in equation (5.9). The obvious reason is that for exact data, the rank of $\mathbf{X}$, and hence $\mathbf{X}^T\mathbf{X}$, is two for $n > 2$. The singularity problem does not arise when data are noisy. Nevertheless, $\mathbf{X}^T\mathbf{X}$ can be very badly conditioned at low noise levels.

Testing the Independence of Model Errors

An alternative approach is to check the statistical property of the residual sequence $\{e(k)\}$ defined by

$$e(k) = y(k) - \mathbf{x}^T(k)\hat{\boldsymbol{\theta}} \tag{5.18}$$

Ideally, if the model order is correct and $\hat{\boldsymbol{\theta}} = \boldsymbol{\theta}$, then $e(k)$ is a statistically independent random process. Therefore, the property of $e(k)$ being an independent random process can be used to test the adequacy of the model order.

To implement this test, we simply compute the serial autocorrelation sequence of $e(k)$ by

$$\phi_{ee}(\tau) = \frac{1}{N - \tau} \sum_{k=1}^{N-\tau} e(k)e(k - \tau) \tag{5.19}$$

for a few lags $\tau = 1, 2, 3, \ldots, M$. Then we check that all $\phi_{ee}(\tau \neq 0)$ are reasonably close to zero for an assumed model order. Since in reality $\phi_{ee}(\tau \neq 0)$ is never exactly zero, we should expect it to oscillate around zero values instead. One helpful way to judge the closeness of $\phi_{ee}(\tau)$ to zero is to count the sign changes of $\phi_{ee}(\tau)$, which should be approximately $M/2$ for M values of lags. Figure 5–4 illustrates a typical case of the influence of model order n on the correlation of residuals for a second-order system. We notice that $\phi_{ee}(\tau)$ ceases to improve significantly for model order $n > 2$ where the true order is $n = 2$.

Now we wish to show how to recursively compute $\hat{\boldsymbol{\theta}}$ as the model order increases. Suppose the model order is increased from n to $n + 1$. We can then apply the recursive algorithm in equation (3.44). Arrange the parameters in the new $\boldsymbol{\theta}$ in the following order:

$$\boldsymbol{\theta} = [a_1, b_0, b_1; a_2, b_2; \ldots; a_n, b_n; \vdots a_{n+1}, b_{n+1}]^T$$

$$= [\boldsymbol{\theta}_{(1)}^T, \boldsymbol{\theta}_{(2)}^T]^T$$

The corresponding matrices of $\mathbf{X}_1, \mathbf{X}_2$ are (with $m > n$)

$$\mathbf{X}_1 = \begin{bmatrix} -y(m), & u(m+1) & u(m); & -y(m-1), & u(m-1) & \ldots; & -y(m-n+1), & u(m-n+1) \\ \vdots \\ -y(m+N-1), & u(m+N), & u(m+N-1); & -y(m+N-2), & u(m+N-2); & \ldots; & -y(m+N-n), & u(m+N-n) \end{bmatrix}$$

Table 5–2
Variations of J as a Function of Model Order n for Different Levels of Noise σ

model order noise level	$n = 1$	$n = 2$	$n = 3$
$\sigma = 0.0$	265.863	0.000	—
$\sigma = 0.1$	248.447	0.987	0.983
$\sigma = 0.5$	337.848	24.558	24.451
$\sigma = 1.0$	308.131	99.863	98.698
$\sigma = 5.0$	5131.905	2462.220	2440.245

Source: Adapted from data in K.J. Åström and P. Eykhoff, "System Identification—A Survey," *Automatica*, vol. 7, pp. 123–162, 1971.

$$\mathbf{X}_2 = \begin{bmatrix} -y(m-n), & u(m-n) \\ \vdots & \\ -y(m+N-n-1), & u(m+N-n-1) \end{bmatrix}$$

and $\mathbf{y}$ is

$$\mathbf{y} = [y(m+1), y(m+2), \ldots, y(m+N)]^T$$

Then according to equation (3.44), we have $\hat{\boldsymbol{\theta}} = [\hat{\boldsymbol{\theta}}_{(1)}^T, \hat{\boldsymbol{\theta}}_{(2)}^T]^T$:

$$\hat{\boldsymbol{\theta}}_{(1)} = \hat{\hat{\boldsymbol{\theta}}}_{(1)} - \mathbf{A}\mathbf{X}_2^T(\mathbf{y} - \mathbf{X}_1\hat{\hat{\boldsymbol{\theta}}}_{(1)})$$

$$\hat{\boldsymbol{\theta}}_{(2)} = \mathbf{B}\mathbf{X}_2^T(\mathbf{y} - \mathbf{X}_1\hat{\hat{\boldsymbol{\theta}}}_{(1)})$$

where

$$\mathbf{A} = (\mathbf{X}_1^T\mathbf{X}_1)^{-1}\mathbf{X}_1^T\mathbf{X}_2\mathbf{B}$$

$$\mathbf{B} = [\mathbf{X}_2^T\mathbf{X}_2 - \mathbf{X}_2^T\mathbf{X}_1(\mathbf{X}_1^T\mathbf{X}_1)^{-1}\mathbf{X}_1^T\mathbf{X}_2]^{-1}$$

$$\hat{\hat{\boldsymbol{\theta}}}_{(1)} = (\mathbf{X}_1^T\mathbf{X}_1)^{-1}\mathbf{X}_1^T\mathbf{y}$$

$\hat{\hat{\boldsymbol{\theta}}}_{(1)}$ is the parameter estimation when the model order is n. The matrix inversion term $(\mathbf{X}_1^T\mathbf{X}_1)^{-1}$ for $\mathbf{A}$ and $\mathbf{B}$ can be updated as follows:

$$\begin{bmatrix} \mathbf{X}_1^T\mathbf{X}_1 & \mathbf{X}_1^T\mathbf{X}_2 \\ \mathbf{X}_2^T\mathbf{X}_1 & \mathbf{X}_2^T\mathbf{X}_2 \end{bmatrix}^{-1} = \begin{bmatrix} \mathbf{C} - \mathbf{A}\mathbf{X}_2^T\mathbf{X}_1\mathbf{C} & -\mathbf{A} \\ -\mathbf{A}^T & \mathbf{B} \end{bmatrix}$$

where

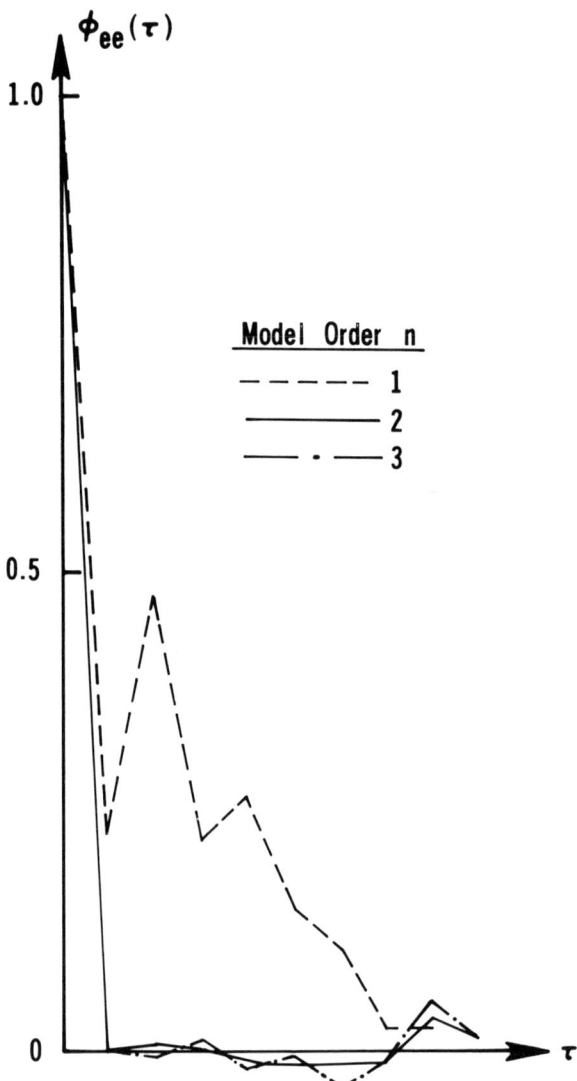

Figure 5–4. A Typical Case of $\phi_{ee}(\tau)$ Illustrating the Influence of Model Order n on the Correlation of Residuals for a Second-order System; $n = 2$ is the Correct Model Order

$$\mathbf{C} = (\mathbf{X}_1^T \mathbf{X}_1)^{-1}.$$

As mentioned in chapter 3, the above algorithm allows us to compute the new $(n + 1)$th order parameter estimates $\hat{\boldsymbol{\theta}} = [\hat{\boldsymbol{\theta}}_{(1)}^T, \hat{\boldsymbol{\theta}}_{(2)}^T]^T$ by simply updating the old nth order parameter estimates $\hat{\boldsymbol{\theta}}_{(1)}$ in which only a 2 ×2 matrix inversion (matrix **B**) is needed. Thus computational cost is very much reduced.

Another identification problem closely related to order determination is time-lag estimation. A system with time lag, or dead time, can be represented by the following difference equation:

$$y(k) = -\sum_{i=1}^{n} a_i y(k - i) + \sum_{i=0}^{n} b_i u(k - i - \tau_d)$$

where τ_d is the constant time lag that is constrained to being an integer. It has been shown (Clark 1967; Hsia 1969b) that τ_d can be successfully identified by repeatedly solving the least-squares parameter estimation problem for a given model order n and a sequence of τ_d values ($\tau_d = 1, 2, \ldots$). The best τ_d estimate is the one that yields the smallest value of the error function J (or the most independent residuals) in a manner similar to the case of model order determination. Thus it is possible to combine time lag and order determination in an identification problem by simply iterating the above procedure for a sequence of n values. This identification algorithm is depicted by the block diagram in figure 5–5.

5.7 Real-time Identification

In this section we discuss the application of sequential algorithms for on-line parameter identification. In particular, the real-time least-squares algorithm for tracking time-varying parameters will be examined.

In chapter 3, we have derived two sequential least-squares algorithms, one of which is the real-time algorithm. This algorithm is a generalization of the on-line least squares that can deal more effectively with time-varying system parameters.

The central idea behind the real-time algorithm for time-varying parameter tracking is of employing an exponentially weighted error function J:

$$J = \sum_{i=1}^{N} \lambda^{N-i} e^2(i), \quad 0 < \lambda < 1 \qquad (5.20)$$

This weighting scheme places more emphasis on the recently acquired data so that the current parameter values dominate the error function.

In presenting the real-time algorithm below, we restrict ourselves to systems with known order n. Therefore, only the parameter values (which are allowed to be time-varying) are to be identified.

In view of equation (3.51) and the error function J in equation (5.20), the recursive parameter updating algorithm can be shown as

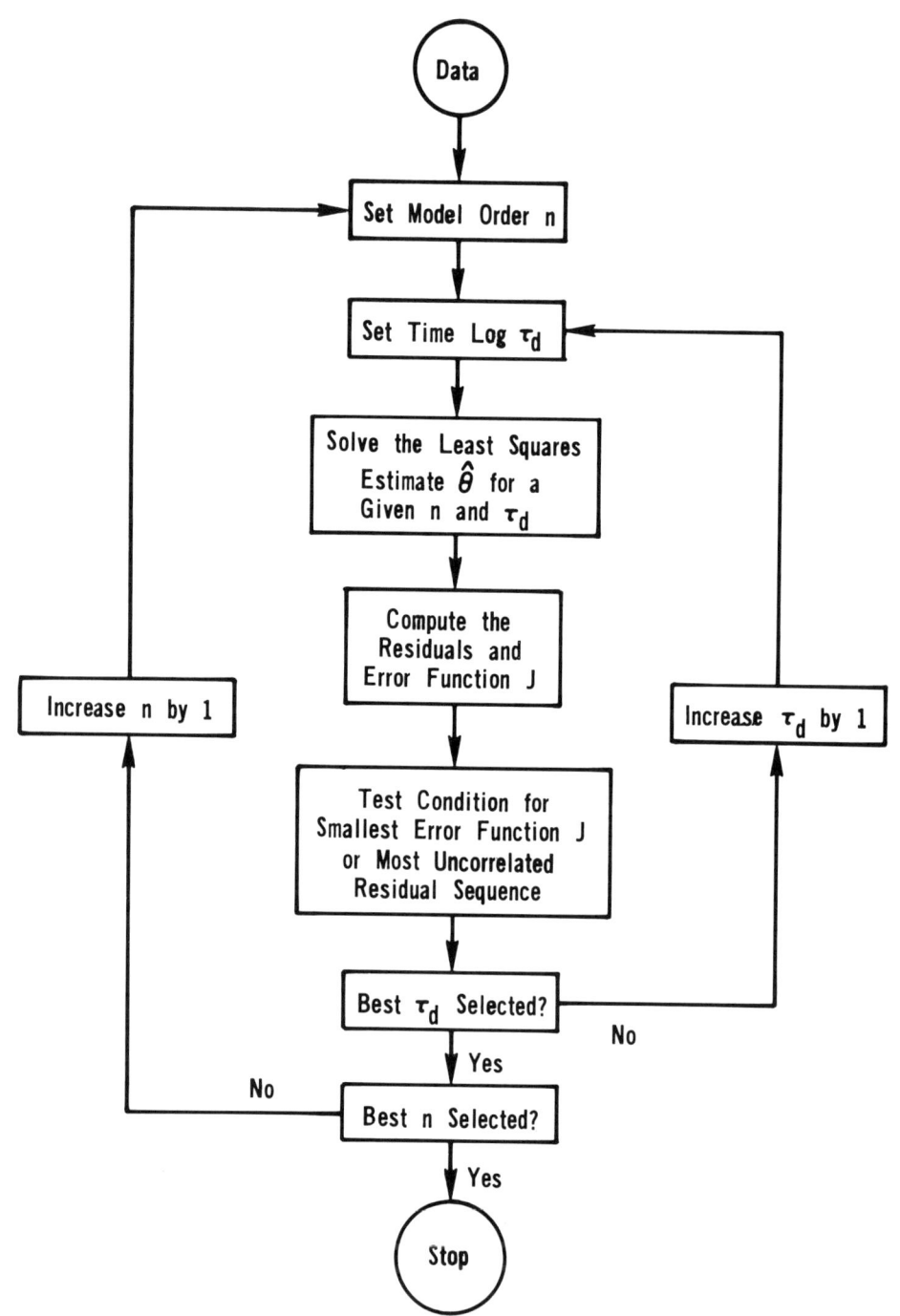

Figure 5–5. Block Diagram Depiction of the Complete System Identification Algorithm

$$\hat{\boldsymbol{\theta}}(N+1) = \hat{\boldsymbol{\theta}}(N) + \gamma(N+1)\mathbf{P}(N)\mathbf{x}(N+1)$$
$$\cdot [y(N+1) - \mathbf{x}^T(N+1)\hat{\boldsymbol{\theta}}(N)]$$

$$\mathbf{P}(N+1) = \frac{1}{\lambda}[\mathbf{P}(N) - \gamma(N+1)\mathbf{P}(N)\mathbf{x}(N+1)\mathbf{x}^T(N+1)\mathbf{P}(N)]$$ (5.21)

$$\gamma(N+1) = 1/[1 + \mathbf{x}^T(N+1)\mathbf{P}(N)\mathbf{x}(N+1)]$$

where

$$\boldsymbol{\theta}(N) = [a_1(N), \ldots, a_n(N), b_0(N), \ldots, b_n(N)]^T$$

$$\mathbf{x}(N+1) = [-y(N), \ldots,$$
$$-y(N+1-n), u(N+1), \ldots, u(N+1-n)]^T$$

$$0 < \lambda < 1.$$

When we let $\lambda = 1$, the above algorithm reduces to that of the usual sequential least squares, which is suitable only for constant parameter systems. In this case, $\mathbf{P}(N) = (\mathbf{X}_N^T\mathbf{X}_N)^{-1}$ and $\lim_{N\to\infty}(\mathbf{X}_N^T\mathbf{X}_N)^{-1} = \mathbf{0}$. This means that the correction gain for updating $\hat{\boldsymbol{\theta}}$ in equation (5.21) is diminishing as N increases, which allows $\hat{\boldsymbol{\theta}}$ to converge to the true constant value. But in the time-varying case, we like to prevent $\mathbf{P}(N)$ from reducing to zero. This is accomplished by introducing the weighting factor λ. In equation (5.21), $\mathbf{P}(N)$ is $\mathbf{P}(N) = (1/\lambda)(\mathbf{X}_N^T\mathbf{X}_N)^{-1}$. For $0 < \lambda < 1$, $\mathbf{P}(N)$ is inflated by the factor $1/\lambda$ which has effectively prevented it from shrinking to zero. Hence, the corresponding algorithm can preserve its updating ability continuously. However, the inherent data truncation effect brought by λ causes increased variances in $\hat{\boldsymbol{\theta}}$ resulting from noise. Thus it is necessary to compromise between fast adaptive capability and the loss of estimate accuracy. In general, λ is chosen close to unity. We now present a simple example to illustrate these properties.

Example [6]

Consider the time-varying system $y(k) + a(k)y(k-1) = b(k)u(k-1) + e(k)$ where $a(k)$ and $b(k)$ have the following values:

$$a(k) = 0.8, \ b(k) = 0.5; \quad \text{for } 0 \leq k < 300$$
$$a(k) = 0.6, \ b(k) = 0.3; \quad \text{for } k \geq 300$$

Here $e(k)$ is a white random variable of zero mean. This system is simulated on the digital computer, and the real-time algorithm is applied to estimate the parameters $\boldsymbol{\theta}(k) = [a(k), b(k)]^T$. The results are shown in figure

5–6 for $\lambda = 0.9$ and $\lambda = 0.99$. We can see that the estimates adapt faster for smaller λ, but at the same time they tend to track the noise more. The opposite effect occurs when λ becomes larger.

Before concluding this section, we wish to point out the relationship of the above algorithm to the well-known *Kalman filtering algorithm*. A detailed discussion of this algorithm is presented in appendix 5A. The central idea of the Kalman filtering algorithm for real-time parameter identification is of modeling the parameter time variations by the randomly driven state transition equation $\boldsymbol{\theta}(k + 1) = \boldsymbol{\theta}(k) + \boldsymbol{\xi}(k)$ in which $\boldsymbol{\xi}(k)$ is a vector of white stationary Gaussian noise with zero mean and covariance matrix $\mathbf{R}$. Coupling this equation with the observation equation $y(k) = \mathbf{x}^T(k)\boldsymbol{\theta}(k) + e(k)$, the following recursive algorithm is derived in appendix 5A:

$$\hat{\boldsymbol{\theta}}(N + 1) = \hat{\boldsymbol{\theta}}(N) + \gamma_1(N + 1)\mathbf{P}_1(N)\mathbf{x}(N + 1)$$
$$\cdot [y(N + 1) - \mathbf{x}^T(N + 1)\hat{\boldsymbol{\theta}}(N)]$$
$$\mathbf{P}_1(N + 1) = \mathbf{P}_1(N) + \mathbf{R}_1 \qquad (5.22)$$
$$- \gamma_1(N + 1)\mathbf{P}_1(N)\mathbf{x}(N + 1)\mathbf{x}^T(N + 1)\mathbf{P}_1(N)$$
$$\gamma_1(N + 1) = 1/[1 + \mathbf{x}^T(N + 1)\mathbf{P}_1(N)\mathbf{x}(N + 1)]$$

in which $\mathbf{R}_1 = \mathbf{R}/\sigma^2$ where σ^2 is the variance of the white stationary Gaussian noise $e(k)$.

We see that $\mathbf{P}_1$ in the above algorithm will never go to zero because of the inclusion of the positive definite covariance matrix $\mathbf{R}_1$. Therefore, the Kalman filtering algorithm can track time-varying parameters. In the absence of exact knowledge of $\mathbf{R}_1$ in practical situations, we usually choose $\mathbf{R}_1$ to be $\mathbf{R}_1 = r\mathbf{I}$ ($r > 0$). The value of r affects the tracking performance in a manner similar to that of the factor $1/\lambda$ in the least-squares algorithm.

5.8 Continuous Systems Identification

Thus far, we have discussed only the identification of discrete system models. In chapter 4, we have shown that the weighting function of a continuous system can be estimated in terms of its sampled values via an equivalent discrete model. However, the most useful representation of continuous systems is the linear differential equation. One obvious approach to parameter identification in differential equation models is to take advantage of the discrete model techniques by first identifying the equivalent difference equations and then calculating the desired parameter values through Z-transformation relationships (Smith 1968; Hsia 1972). However, this approach has severe limitations.

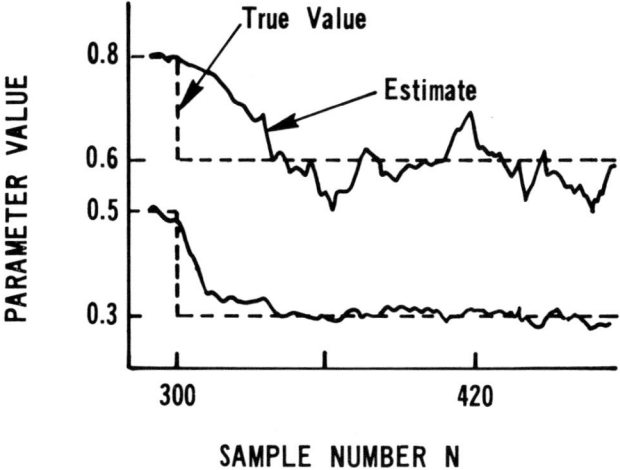

Figure 5–6a. Real-time Identification Results of a First-order System with Noise-corrupted Data and $\lambda = 0.90$

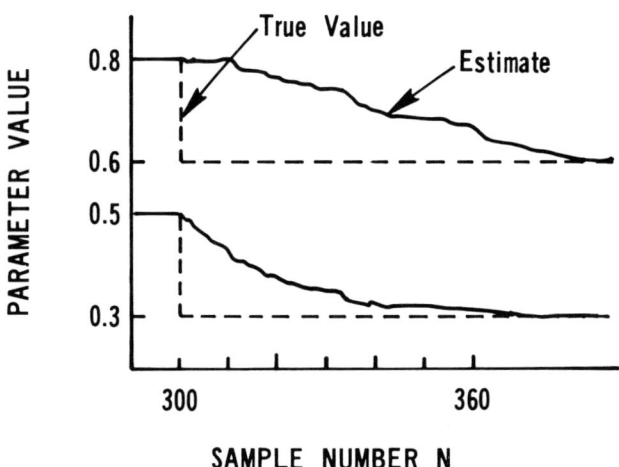

Figure 5–6b. Real-time Identification Results of a First-order System with Noise-free Data and $\lambda = 0.90$.

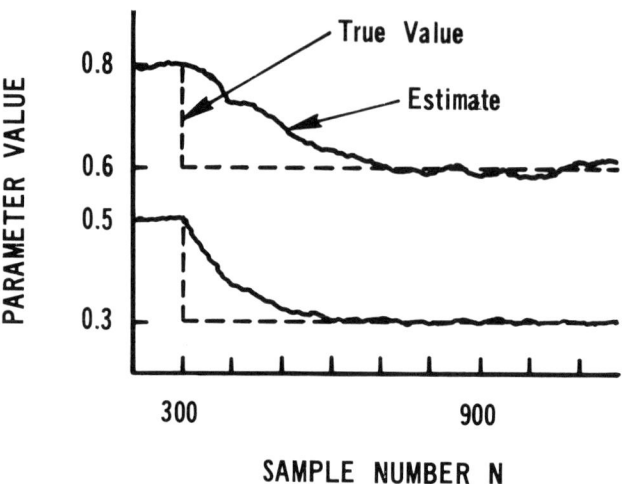

Figure 5–6c. Real-time Identification Results of a First-order System with Noise-corrupted Data and $\lambda = 0.99$

Although discrete approximation models of continuous systems are useful in many engineering applications, differential equation models are often desired in other areas of study, such as the life sciences, economics, biomedicine, and so on. In these areas, the coefficients in the differential equation, which describe the nature of the physical or chemical processes occuring inside the system, have direct relationships to the system's physical or chemical parameters that we wish to identify. In these situations, we need to directly identify the differential equation model at hand.

From the viewpoint of system identification, there is a fundamental difference between a differential equation and a difference equation. In the latter, the system data are the past and present input-output signals that can be easily measured for use in identification. On the other hand, system data appearing in a differential equation are time derivatives of the input-output signals that are normally inaccessible (or impractical) for measurement. Thus some form of reconstruction of the time derivatives has to be implemented. Such a difficulty has made differential equation models more cumbersome to identify. In this section, we present an identification method in which the least-squares theory is employed.

A linear time-invariant nth order system can be described by the following differential equation

$$\frac{d^n}{dt^n}y(t) + a_{n-1}\frac{d^{n-1}}{dt^{n-1}}y(t) + \ldots + a_0 y(t)$$

(5.23)

$$= b_{n-1}\frac{d^{n-1}}{dt^{n-1}}u(t) + \ldots + b_0 u(t)$$

This equation can be simplified as $A(p)y(t) = B(p)u(t)$ where p is the differential operator defined by $p = d/dt$, and

$$A(p) = p^n + a_{n-1}p^{n-1} + \ldots + a_0$$

$$B(p) = b_{n-1}p^{n-1} + \ldots + b_0$$

As shown in chapter 2, the corresponding system transfer function is $H(s) = B(s)/A(s)$.

Define the time derivatives at $t = t_k$ as

$$y_i(k) = \frac{d^i}{dt^i}y(t_k) \quad u_i(k) = \frac{d^i}{dt^i}u(t_k)$$

Then equation (5.23) at $t = t_k$ becomes

$$y_n(k) + \sum_{i=0}^{n-1} a_i y_i(k) = \sum_{i=0}^{n-1} b_i u_i(k) \quad (5.24)$$

Suppose we define

$$\boldsymbol{\theta} = [a_0, \ldots, a_{n-1}, b_0, \ldots, b_{n-1}]^T$$

$$\mathbf{x}(k) = [-y_0(k), \ldots, -y_{n-1}(k), u_0(k), \ldots, u_{n-1}(k)]^T$$

Then the system equation can be formulated as

$$y_n(k) = \mathbf{x}^T(k)\boldsymbol{\theta} + e(k) \quad (5.25)$$

in which the error term $e(k)$ is introduced to account for random modeling error. As before, we assume that $e(k)$ is a zero-mean white random variable. A block diagram representation of the modeling problem is shown in figure 5-7.

The above formulation suggests that the parameter vector $\boldsymbol{\theta}$ can be estimated by means of least-square error if the derivative signals $y_i(k)$ and $u_i(k)$ are directly measurable. Let us suppose such measurements are made for $k = 1, 2, \ldots, N$. Then we can set up the vector equation

$$\mathbf{y} = \mathbf{X}\boldsymbol{\theta} + \mathbf{e} \quad (5.26)$$

where

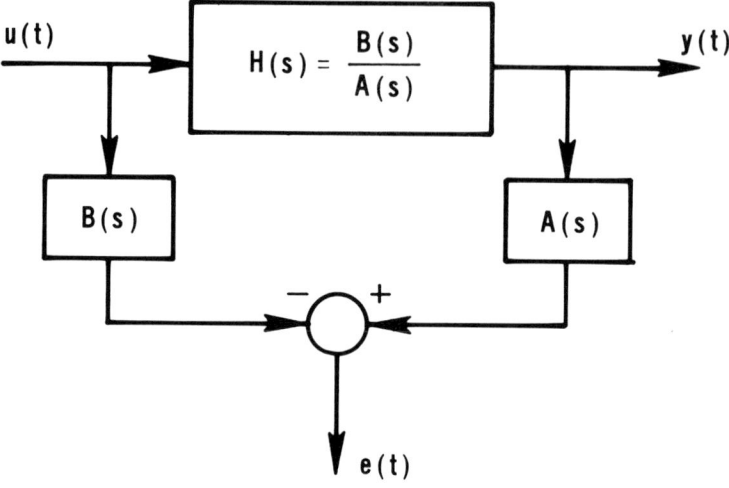

Figure 5-7. Modeling Configuration of Continuous Systems; $e(t)$ is the Equation Error

$$\mathbf{y} = [y_n(1), \ldots, y_n(N)]^T$$
$$\mathbf{e} = [e(1), \ldots, e(N)]^T$$
$$\mathbf{X} = [\mathbf{x}(1), \ldots, \mathbf{x}(N)]^T$$

Applying the least-squares procedure to equation (5.26) yields the parameter vector estimate $\hat{\boldsymbol{\theta}}$:

$$\hat{\boldsymbol{\theta}} = (\mathbf{X}^T\mathbf{X})^{-1}\mathbf{X}^T\mathbf{y} \tag{5.27}$$

In practice, this solution is nonrealistic because the derivative signals are seldom available. Although ideally they can be obtained by successive differentiation of the input and output signals, ideal differentiators are physically unrealizable. Therefore, we have to substitute for them with realizable filters. In the following discussion, the concept of *state variable filters* is introduced (Lion 1966).

Suppose we take the Laplace transform of equation (5.23) and, introducing the filter $\mathbf{T}(s)$, we get the steady state equation

$$(s^n + a_{n-1}s^{n-1} + \ldots + a_0)\mathbf{T}(s)Y(s) = (b_{n-1}s^{n-1} + \ldots + b_0)\mathbf{T}(s)U(s) \tag{5.28}$$

Define a set of new filters $M_i(s)$:

$$M_i(s) = \mathbf{T}(s)s^i \quad i = 1, 2, \ldots, n \tag{5.29}$$

and the corresponding filtered signals $\tilde{Y}_i(s)$ and $\tilde{U}_j(s)$:

$$\tilde{Y}_i(s) = M_i(s)Y(s), \quad i = 0, 1, \ldots, n$$

$$\tilde{U}_j(s) = M_j(s)U(s), \quad j = 0, 1, \ldots, n-1$$

Then equation (5.28) yields a filtered equation

$$\tilde{Y}_n(s) + a_{n-1}\tilde{Y}_{n-1}(s) + \ldots + a_0\tilde{Y}_0(s) = b_{n-1}\tilde{U}_{n-1}(s) + \ldots + b_0\tilde{U}_0(s)$$

This provides the following time domain equation:

$$\tilde{y}_n(t) + \sum_{i=0}^{n-1} a_i \tilde{y}_i(t) = \sum_{i=0}^{n-1} b_i \tilde{u}_i(t) \tag{5.30}$$

which is similar to equation (5.24). Thus the parameters can be estimated by

$$\hat{\boldsymbol{\theta}} = (\tilde{\mathbf{X}}^T \tilde{\mathbf{X}})^{-1} \tilde{\mathbf{X}}^T \tilde{\mathbf{y}} \tag{5.31}$$

where $\tilde{\mathbf{X}}$ and $\tilde{\mathbf{y}}$ are similarly defined as $\mathbf{X}$ and $\mathbf{y}$ in equation (5.26).

The condition of this solution procedure is that the filters $M_i(s)$ are physically realizable. We can easily fulfill this condition by choosing an appropriate $T(s)$. We refer to $M_i(s)$ as *state variable filters*. The block diagram in figure 5-8 depicts such a filtering system.

What we have shown is that continuous systems can be identified in a manner similar to the discrete system provided that state variable filters are employed and sampling of the filtered signals is introduced. The parameter estimates so obtained enjoy all the statistical properties guaranteed by least-squares theory. It is of course also possible to process the continuous signals for parameter estimation using on-line steepest decent (Lion 1966) and model follower (Hsia 1969a) techniques.

5.9 Concluding Remarks

In this chapter, we have introduced the very basic results of identification of linear difference equation models. We have formulated the system identification problem so that the simple least-squares theory is readily applicable. The topics discussed include parameter estimation, structure determination, and on-line and real-time identification. Also discussed at length was the identification of continuous systems modeled by differential equations. It was demonstrated that all the least-squares estimation algorithms are applicable to this case when the input-output signals are filtered and sampled.

However, these results are suitable only for systems in which the noise disturbances are small, whereby the fitting error is essentially a

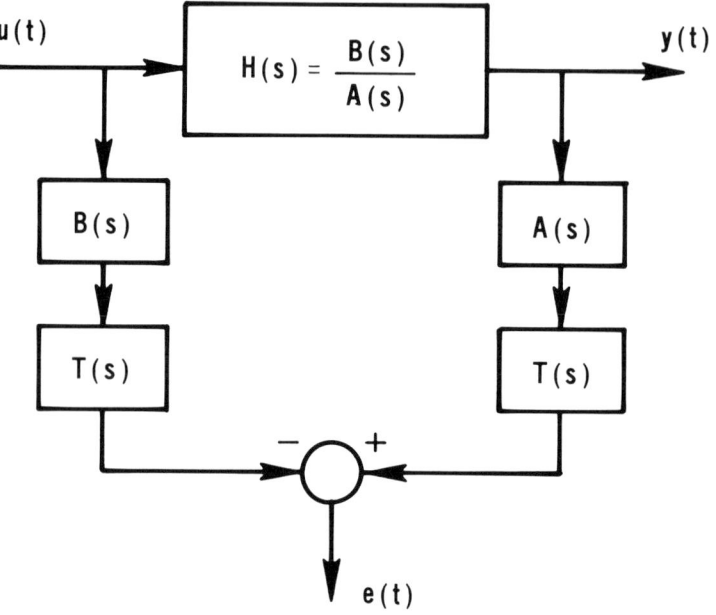

Figure 5–8. Modeling Configuration Using State Variable Filters

white random process. This condition fails when strong noise disturbances are present. In those cases, the method of generalized least squares will have to be employed in order to get good parameter estimates. This problem is discussed in the next chapter.

References

Åström, K. J., and Bohlin, T., "Numerical Identification of Linear Dynamic Systems from Normal Operating Records," *IFAC Symposium—Theory of Self-adaptive Control Systems*, Teddington, England, 1966.

Åström, K. J., and Eykhoff, P., "System Identification—a Survey," *Automatica*, Vol. 7, pp. 123–162, 1971.

Clark, D. W., "Generalised-least-squares Estimation of Parameters of a Dynamic Model," *IFAC Symposium—Identification in Automatic Control Systems*, Paper 3.17, Prague, 1967.

Hastings-James, R., and Sage, M. W., "Recursive Generalised-least-squares Procedures for On-line Identification of Process Parameters," *Proceedings IEEE*, Vol. 116, pp. 2057–2062, 1969.

Hsia, T.C., "An On-line Technique for System Identification," *IEEE Transactions on Automatic Control*, Vol. AC-14, pp. 92–96, Feb. 1969a.

_____, "A Discrete Method for Parameter Identification in Linear Systems with Transport Lags," *IEEE Transactions on Aerospace and Electronic Systems*, Vol. AES-5, pp. 236–239, Mar. 1969b.

_____, "On Sampled Data Approach to Parameter Identification of Continuous Linear Systems," *IEEE Transactions on Automatic Control*, Vol. AC-17, pp. 247–249, Apr. 1972.

Kalman, R.E., "Design of a Self-optimizing Control System," *Transactions ASME*, Vol. 80, pp. 468–478, Feb. 1958.

Lion, P.M., "Rapid Identification of Linear and Nonlinear Systems," *JACC Proceedings*, pp. 605–614, 1966.

Smith, F.W., "System Laplace-transform Estimation from Sampled Data," *IEEE Transactions on Automatic Control*, Vol. AC-13, pp. 37–44, Feb. 1968.

Unbehanen, H., and Göbring, B., "Tests for Determining Order in Parameter Estimation," *Automatica*, Vol. 10, pp. 233–244, May 1974.

Appendix 5A
Kalman Filtering Approach to System Parameter Identification

An interesting approach to recursive parameter estimation is to apply the Kalman (1960) filtering theory. This well-known theory in the field of optimal estimation has a very close relationship to our recursive least-squares algorithm. More significantly, the theory is formulated in terms of time-varying systems. This means that it is possible to obtain from this theory a rigorous real-time parameter identification algorithm (Wieslander 1969; Wieslander and Wittenmark 1971). Here we present the Kalman filtering formulation of the system parameter estimation problem.

First, we introduce Kalman's result. Consider a time-varying linear discrete-time system in state variable form

$$\omega(k + 1) = \Phi(k + 1)\omega(k) + \Gamma(k + 1)\xi(k)$$

$$\mathbf{y}(k) = \mathbf{H}(k)\omega(k) + \mathbf{e}(k)$$

(5A.1)

where ω is the n state vector, $\mathbf{y}$ is the m output vector, Φ is the state transition matrix, Γ is the disturbance transition matrix, $\mathbf{H}$ is the output matrix, and $\xi(k)$ and $\mathbf{e}(k)$ are vectors of Gaussian independent random variables with zero means and covariance matrices $\mathbf{R}(k)$ and $\Sigma(k)$ respectively.

KALMAN FILTERING THEOREM. Let the initial condition $\omega(0)$ be a Gaussian random vector with mean $\mathbf{m}_0$ and covariance matrix $\mathbf{C}_0$. The best estimate of $\omega(k + 1)$, in the sense of minimum variance, given the observed outputs $\mathbf{y}(1), \mathbf{y}(2), \ldots, \mathbf{y}(k + 1)$, is given by the recursive formula

$$\hat{\omega}(k + 1) = \Phi(k + 1)\hat{\omega}(k) + \mathbf{G}(k + 1)$$
$$\cdot [\mathbf{y}(k + 1) - \mathbf{H}(k + 1)\Phi(k + 1)\hat{\omega}(k)] \quad (5A.2)$$

where

$$\mathbf{G}(k + 1) = \mathbf{S}(k + 1)\mathbf{H}^T(k + 1)$$
$$\cdot [\mathbf{H}(k + 1)\mathbf{S}(k + 1)\mathbf{H}^T(k + 1) + \Sigma(k + 1)]^{-1}$$

$$\mathbf{S}(k + 1) = \Phi(k + 1)\mathbf{P}(k)\Phi^T(k + 1)$$
$$+ \Gamma(k + 1)\mathbf{R}(k)\Gamma^T(k + 1)$$

$$\mathbf{P}(k + 1) = \mathbf{S}(k + 1) - \mathbf{G}(k + 1)\mathbf{H}(k + 1)\mathbf{S}(k + 1)$$

The initial conditions are $\hat{\boldsymbol{\omega}}(0) = \mathbf{m}_0$ and $\mathbf{S}(0) = \mathbf{C}_0$.

Now we will apply this result to the parameter identification problem of chapter 5. Let us assume that the system parameters are time-varying and the variations can be modeled by the relationship (a randomly driven time-varying linear system)

$$\boldsymbol{\theta}(k + 1) = \boldsymbol{\Phi}(k + 1)\boldsymbol{\theta}(k) + \boldsymbol{\xi}(k) \tag{5A.3}$$

Here the transition matrix $\boldsymbol{\Phi}(k + 1)$ describes the deterministic portion of the variation and the random vector $\boldsymbol{\xi}(k)$ accounts for the parameter perturbations that are unpredictable. Coupling this equation with the scalar system model equation developed in chapter 5,

$$y(k) = \mathbf{x}^T(k)\boldsymbol{\theta}(k) + e(k) \tag{5A.4}$$

we have a set of equations analogous to equation (5A.1). Therefore, the algorithms in equation (5A.2) can be readily applied to estimate the time-varying parameters $\boldsymbol{\theta}(k)$ when the following identification of variables are made:

$$\boldsymbol{\omega} = \boldsymbol{\theta}, \qquad \boldsymbol{\Gamma}(k) = \mathbf{I}, \qquad \mathbf{H}(k) = \mathbf{x}^T(k)$$

However, since the resulting algorithm would be very complicated, we will instead consider a simplified case for which the results will be compatible with the least-squares algorithm of chapter 5.

Consider that $\boldsymbol{\Phi}(k) = \mathbf{I}$ and $\boldsymbol{\Gamma}(k) = \mathbf{I}$ in equation (5A.3). This yields

$$\boldsymbol{\theta}(k + 1) = \boldsymbol{\theta}(k) + \boldsymbol{\xi}(k) \tag{5A.5}$$

Furthermore, assume that $\boldsymbol{\xi}(k)$ is a stationary random process so that its covariance matrix $\mathbf{R}$ is constant. Because $e(k)$ in equation (5A.4) is a scalar, and we also assume it to be a stationary random process, its covariance matrix becomes $\boldsymbol{\Sigma}(k) = \sigma^2$.

Now let $\boldsymbol{\Phi}(k) = \mathbf{I}, \boldsymbol{\Gamma}(k) = \mathbf{I}, \mathbf{H}(k) = \mathbf{x}^T(k), \mathbf{R}(k) = \mathbf{R}$, and $\boldsymbol{\Sigma}(k) = \sigma^2$. The algorithm in equation (5A.2), when applied to the system of equations (5A.4) and (5A.5), becomes (after simplification)

$$\hat{\boldsymbol{\theta}}(k + 1) = \hat{\boldsymbol{\theta}}(k) + \gamma(k + 1)\mathbf{P}(k)\mathbf{x}(k + 1)\,[y(k + 1) - \mathbf{x}^T(k + 1)\hat{\boldsymbol{\theta}}(k)]$$

$$\mathbf{P}(k + 1) = \mathbf{P}(k) + \mathbf{R} - \gamma(k + 1)\mathbf{P}(k)\mathbf{x}(k + 1)\mathbf{x}^T(k + 1)\mathbf{P}(k) \tag{5A.6}$$

$$\gamma(k + 1) = 1/[\sigma^2 + \mathbf{x}^T(k + 1)\mathbf{P}(k)\mathbf{x}(k + 1)]$$

1. To apply the above algorithm, we must know the covariance matrix $\mathbf{R}$ and variance σ^2 a priori. Any approximation of $\mathbf{R}$ and σ^2 would render the algorithm suboptimal.

2. If we define $\mathbf{R}_1 = \mathbf{R}/\sigma^2$, $\mathbf{P}_1 = \mathbf{P}/\sigma^2$, and $\gamma_1 = \sigma^2\gamma$, we get a scaled algorithm as follows:

$$\hat{\boldsymbol{\theta}}(k+1) = \hat{\boldsymbol{\theta}}(k) + \gamma_1(k+1)\mathbf{P}_1(k)\mathbf{x}(k+1)\left[y(k+1) - \mathbf{x}^T(k+1)\hat{\boldsymbol{\theta}}(k)\right]$$

$$\mathbf{P}_1(k+1) = \mathbf{P}_1(k) + \mathbf{R}_1 - \gamma_1(k+1)\mathbf{P}_1(k)\mathbf{x}(k+1)\mathbf{x}^T(k+1)\mathbf{P}_1(k) \quad (5A.7)$$

$$\gamma_1(k+1) = 1/[1 + \mathbf{x}^T(k+1)\mathbf{P}_1(k)\mathbf{x}(k+1)]$$

3. When we let $\boldsymbol{\xi}(k) = \mathbf{0}$ in equation (5A.5), we have a constant parameter model $\boldsymbol{\theta}(k+1) = \boldsymbol{\theta}(k)$. In this case $\mathbf{R}_1 = \mathbf{0}$ in equation (5A.7), and the algorithm becomes identical to the least-squares algorithm of equation (3.31).

4. For the case where $\boldsymbol{\theta}(k)$ only varies deterministically with known variational characteristics, the parameter dynamic model is

$$\boldsymbol{\theta}(k+1) = \boldsymbol{\Phi}(k+1)\boldsymbol{\theta}(k) \quad (5A.8)$$

where $\boldsymbol{\Phi}$ is the appropriate transition matrix. Experimental studies have shown that excellent tracking performance can be obtained by this approach (Lee 1964).

References

Kalman, R. E., "A New Approach to Linear Filtering and Prediction Problems," *Journal of Basic Engineering*, Vol. 82D, pp. 35–45, March, 1960.

Lee, R. C. K., *Optimal Estimation, Identification, and Control*, MIT Press, Cambridge, Mass., 1964.

Wieslander, J., "Real-time Identification, Part II," *Report 6909*, Lund Institute of Technology, Division of Automatic Control, 1969.

Wieslander, J., and Wittenmark, B., "An Approach to Adaptive Control Using Real-time Identification," *Automatica*, Vol. 7, pp. 211–217, 1971.

6

System Identification by Generalized Least Squares

6.1 Introduction

In this chapter, we introduce the method of generalized least squares for system parameter identification. This method is more powerful than the least-squares method introduced in the last chapter in that it can provide more accurate estimates when the residuals (or equation errors) are autocorrelated. Correlated residuals arise in system identification when random disturbances are present in the system; application of the least-squares method would yield biased estimates. We explain why bias occurs and how the generalized least-squares approach can effectively deal with the bias problem. In addition, we briefly review the method of instrumental variables as an alternative approach to obtaining consistent parameter estimates.

6.2 Formulation of Noisy System Model

The noisy system model being considered here is shown in figure 6–1. The additive noise $v(k)$ accounts for the random disturbances in the system as well as for measurement errors. We assume that $v(k)$ is a stationary random process with zero mean and autocorrelation function $R_{vv}(\tau)$. It is also assumed that $v(k)$ is uncorrelated with either the input $u(k)$ or the output $w(k)$. The observable signals are $u(k)$ and $y(k)$.

As in chapter 5, we let the discrete system be described by the difference equation

$$A(q^{-1})w(k) = B(q^{-1})u(k)$$

$$y(k) = w(k) + v(k)$$

(6.1)

where q^{-1} is the backward shifting operator and

$$A(q^{-1}) = 1 + a_1 q^{-1} + \ldots + a_n q^{-n}$$

$$B(q^{-1}) = b_0 + b_1 q^{-1} + \ldots + b_n q^{-n}$$

We assume, as before, that the system is stable, the order n is known a priori, and the input is persistently exciting of order $2n + 1$. Reducing equation (6.1) yields the following noisy system model:

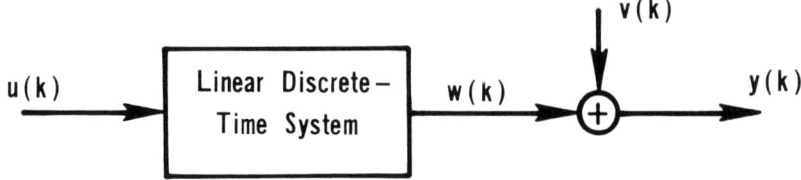

Figure 6-1. Noisy System Model Configuration

$$A(q^{-1})y(k) = B(q^{-1})u(k) + A(q^{-1})v(k) \tag{6.2}$$

Defining the residual term $\epsilon(k)$ as

$$\epsilon(k) = A(q^{-1})v(k) \tag{6.3}$$

we have the familiar form

$$A(q^{-1})y(k) = B(q^{-1})u(k) + \epsilon(k) \tag{6.4}$$

It is clear that $\epsilon(k)$ is in general an autocorrelated random process. We refer here to $\epsilon(k)$ as the *correlated residual*. In chapter 5, the residual $e(k)$, as in equation (5.4) is called the *white* (or *uncorrelated*) *residual*. The present chapter is essentially devoted to the problem of biases in least-squares estimates resulting from correlated residuals.

6.3 Bias Problems Associated with Correlated Residuals

In this section, we show that bias arises when applying the least-squares estimation to the noisy model in equation (6.4). We first set up the least-squares solution of the parameter estimates $\hat{\theta}$

$$\hat{\theta} = (X^T X)^{-1} X^T y$$

where

$$X = \begin{bmatrix} -y(n), & \ldots, & -y(1) & \vdots & u(n+1), & \ldots, & u(1) \\ -y(n+1), & \ldots, & -y(2) & \vdots & u(n+2), & \ldots, & u(2) \\ \vdots & & \vdots & \vdots & \vdots & & \\ -y(n+N-1), & \ldots, & -y(N) & \vdots & u(n+N), & \ldots, & u(N) \end{bmatrix} \tag{6.5}$$

$$y = [y(n+1), y(n+2), \ldots, y(n+N)]^T$$

$$\theta = [a_1, \ldots, a_n, b_0, \ldots, b_n]^T$$

Based on the analysis in chapter 3, we can express the expectation of $\hat{\boldsymbol{\theta}}$ as

$$E[\hat{\boldsymbol{\theta}}] = \boldsymbol{\theta} + E[(\mathbf{X}^T\mathbf{X})^{-1}\mathbf{X}^T\boldsymbol{\epsilon}] \quad (6.6)$$

where $\boldsymbol{\epsilon}$ is the residual vector defined by

$$\boldsymbol{\epsilon} = [\epsilon(n+1), \epsilon(n+2), \ldots, \epsilon(n+N)]^T$$

It is seen that $\hat{\boldsymbol{\theta}}$ is biased if

$$E[(\mathbf{X}^T\mathbf{X})^{-1}\mathbf{X}^T\boldsymbol{\epsilon}] \neq \mathbf{0}, \quad \text{or simply}$$

$$E[\mathbf{X}^T\boldsymbol{\epsilon}] \neq 0 \quad (6.7)$$

We now demonstrate that the above condition is true for the noisy model in equation (6.4). Since $y(k) = w(k) + v(k)$, we can partition the matrix $\mathbf{X}$ in equation (6.5) into two parts:

$$\mathbf{X}_{y,u} = \mathbf{X}_{w,u} + \mathbf{X}_{v,0}$$

where, specifically,

$$\mathbf{X}_{v,0} = \begin{bmatrix} -v(n), & \ldots, & -v(1) & \vdots \\ & \vdots & & \vdots & 0 \\ -v(n+N-1), & \ldots, & -v(N) & \vdots \end{bmatrix}$$

Since $E[\mathbf{X}_{w,u}^T\boldsymbol{\epsilon}] = \mathbf{0}$, we only have to show that $E[\mathbf{X}_{v,0}^T\boldsymbol{\epsilon}] \neq \mathbf{0}$. It is easy to see that the elements in $E[\mathbf{X}_{v,0}^T\boldsymbol{\epsilon}]$ are

$$E[v(i)\epsilon(j)] = R_{v\epsilon}(\tau), \quad \tau = 1, 2, \ldots, n$$

In view of equation (6.3), we have

$$\epsilon(k) = v(k) + \sum_{i=1}^{n} a_i v(k-i) \quad (6.8)$$

Hence

$$R_{v\epsilon}(\tau) = R_{vv}(\tau) + \sum_{i=1}^{n} a_i R_{vv}(\tau - i)$$

As a result of the earlier assumption that $v(k)$ is autocorrelated and $R_{vv}(\tau) \neq 0$ for all τ, we can conclude that not all $R_{v\epsilon}(\tau) = 0$. Hence $E[\mathbf{X}^T\boldsymbol{\epsilon}] \neq \mathbf{0}$, and $\hat{\boldsymbol{\theta}}$ is biased. The only exception to this general conclusion is when $R_{vv}(\tau)$ satisfies the condition

$$R_{vv}(\tau) + \sum_{i=1}^{n} a_i R_{vv}(\tau - i) = 0, \quad \tau = 1, 2, \ldots, n$$

This condition suggests that $v(k)$ satisfies the difference equation

$$v(k) + \sum_{i=1}^{n} a_i v(k - i) = e(k) \tag{6.9}$$

or

$$A(q^{-1})v(k) = e(k) \tag{6.10}$$

where $e(k)$ is a white random noise. Comparing equations (6.8) and (6.9) yields the relationship $\epsilon(k) = e(k)$, which means that the system equation (6.4) has white residual. In this special case, the estimate $\hat{\theta}$ is unbiased as expected. Equation (6.10) implies that the random $v(k)$ is a filtered white noise with filter transfer function

$$\frac{v(z)}{e(z)} = \frac{1}{A(z^{-1})} \tag{6.11}$$

This special case has been pointed out in figure 5.3.

6.4 Formulation of the Generalized Least-squares Problem

Having shown the existence of bias, we can appreciate the need to modify the least-squares solution in order to achieve better parameter identification. This need has led to the development of the method of generalized least squares (GLS) (Clarke 1967). For the sake of clarity, we denote the method of least squares of chapter 5 as LS. It will become apparent later that LS is a special case of GLS.

The key consideration in using the GLS approach is that it introduces a whitening filter to convert the correlated residual into a white residual. To implement this idea, we assume that the residual $\epsilon(k)$ has a rational power spectrum so that the following autoregressive model is satisfied:

$$\epsilon(k) + \sum_{i=1}^{p} c_i \epsilon(k - i) = e(k) \tag{6.12}$$

in which c_i are constant coefficients and p is the order of the model. In general, c_i and p are unknown a priori. Practical experience has shown that a good model can be obtained by preassigning p as 2 or 3, and then optimizing c_i. Rewrite equation (6.12) as

$$C(q^{-1})\epsilon(k) = e(k) \tag{6.13}$$

with

$$C(q^{-1}) = 1 + c_1 q^{-1} + \ldots + c_p q^{-p}$$

This implies that $\epsilon(k)$ is a filtered white noise with filter transfer function

$$\frac{\epsilon(z)}{e(z)} = \frac{1}{C(z^{-1})} \qquad (6.14)$$

Combining equations (6.3) and (6.13), we get the transfer function

$$\frac{v(z)}{e(z)} = \frac{1}{A(z^{-1})C(z^{-1})} \qquad (6.15)$$

A block diagram showing the system and noise transfer function is given in figure 6–2. Finally, we can derive the following system-noise model by combining equations (6.4) and (6.13):

$$C(q^{-1})A(q^{-1})y(k) = C(q^{-1})B(q^{-1})u(k) + e(k) \qquad (6.16)$$

This model now has a white residual $e(k)$. Consistent parameter estimates of a_i b_i c_i can be obtained by minimizing the following error function:

$$J = \sum_k e^2(k) = \sum_k [C(q^{-1})A(q^{-1})y(k) - C(q^{-1})B(q^{-1})u(k)]^2 \qquad (6.17)$$

In the context of system identification, $e(k)$ is called the *generalized equation error*. The approach to parameter estimation of minimizing the above generalized error function J is called the *GLS method*. The filter $C(z^{-1})$ is called the *whitening filter*. These terms are illustrated in figure 6–3.

6.5 Generalized Least-squares Estimation Algorithm

In this section, the algorithm of GLS parameter estimation is derived. It is clear from equation (6.16) that to estimate the system parameters a_i and b_i, it is necessary also to estimate the residual autoregression coefficients c_i. However, equation (6.16) is not linear in the polynomials $A(q^{-1})$, $B(q^{-1})$, and $C(q^{-1})$. Therefore, these parameters cannot be estimated by linear procedures. Instead, a numerical procedure is called for. The GLS algorithm is essentially a relaxation method in which J is minimized first with respect to a_i b_i, and then with respect to c_i. This process is then iterated. In each step of minimization, we need only to solve a simple least-squares problem. The GLS algorithm is described below.

Step 1. To start, let $c_i = 0$ or $C(q^{-1}) = 1$. The error function J becomes

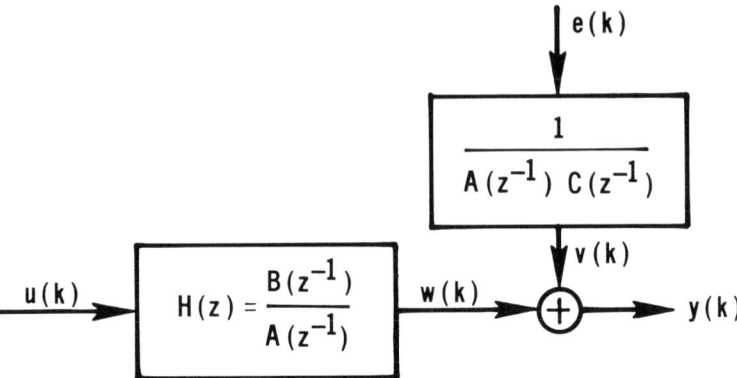

Figure 6-2. Transfer Function Block Diagram of the Noisy System

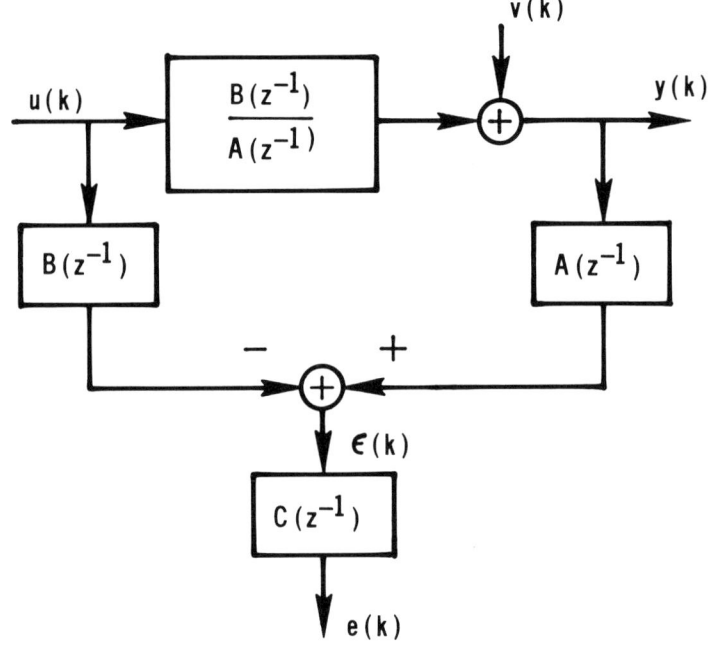

Figure 6-3. Block Diagram Showing the Generalized Equation Error

$$J_1 = \sum_k [A(q^{-1})y(k) - B(q^{-1})u(k)]^2$$
$$= (\mathbf{y} - \mathbf{X}\boldsymbol{\theta})^T(\mathbf{y} - \mathbf{X}\boldsymbol{\theta}) \tag{6.18}$$

Minimizing J_1 with respect to $\boldsymbol{\theta}$ yields

$$\hat{\boldsymbol{\theta}} = (\mathbf{X}^T\mathbf{X})^{-1}\mathbf{X}^T\mathbf{y}$$

Step 2. With $A(q^{-1})$ and $B(q^{-1})$ estimated, define the residual $\epsilon(k)$ by

$$\epsilon(k) = \hat{A}(q^{-1})y(k) - \hat{B}(q^{-1})u(k)$$

Thus J can be written as

$$J_2 = \sum_k [C(q^{-1})\epsilon(k)]^2$$
$$= (\boldsymbol{\epsilon} - \boldsymbol{\Omega}\mathbf{c})^T(\boldsymbol{\epsilon} - \boldsymbol{\Omega}\mathbf{c}) \qquad (6.19)$$

where

$$\mathbf{c} = [c_1, c_2, \ldots, c_p]^T$$
$$\boldsymbol{\epsilon} = [\epsilon(n+1), \epsilon(n+2), \ldots, \epsilon(n+N)]^T$$

$$\boldsymbol{\Omega} = \begin{bmatrix} -\epsilon(n), & -\epsilon(n-1), \ldots, & -\epsilon(n+1-p) \\ -\epsilon(n+1), & -\epsilon(n), \ldots, & -\epsilon(n+2-p) \\ \vdots & & \\ -\epsilon(n+N-1), & \ldots\ldots\ldots, & -\epsilon(n+N-p) \end{bmatrix}$$

The estimate $\hat{\mathbf{c}}$ that minimizes J_2 is given by the standard LS solution

$$\hat{\mathbf{c}} = (\boldsymbol{\Omega}^T\boldsymbol{\Omega})^{-1}\boldsymbol{\Omega}^T\boldsymbol{\epsilon}$$

Step 3. From $\hat{\mathbf{c}}$ above, we define two filtered signals $\tilde{u}(k)$ and $\tilde{y}(k)$ as

$$\tilde{u}(k) = \hat{C}(q^{-1})u(k) \quad \tilde{y}(k) = \hat{C}(q^{-1})y(k)$$

Then J becomes

$$J_3 = \sum_k [A(q^{-1})\tilde{y}(k) - B(q^{-1})\tilde{u}(k)]^2$$
$$= (\tilde{\mathbf{y}} - \tilde{\mathbf{X}}\boldsymbol{\theta})^T(\tilde{\mathbf{y}} - \tilde{\mathbf{X}}\boldsymbol{\theta}) \qquad (6.20)$$

in which $\tilde{\mathbf{y}}$ and $\tilde{\mathbf{X}}$ result from substituting $\tilde{y}(k)$ and $\tilde{u}(k)$ into $\mathbf{y}$ and $\mathbf{X}$. We now have an improved estimate

$$\hat{\boldsymbol{\theta}} = (\tilde{\mathbf{X}}^T\tilde{\mathbf{X}})^{-1}\tilde{\mathbf{X}}^T\tilde{\mathbf{y}}$$

Step 4. Return to step 2. The computations are repeated until $\hat{\boldsymbol{\theta}}$ and $\hat{\mathbf{c}}$ are converged. The above GLS algorithm is illustrated by the flow diagram of figure 6–4. A block diagram interpretation of the GLS identification scheme is given in figure 6–5.

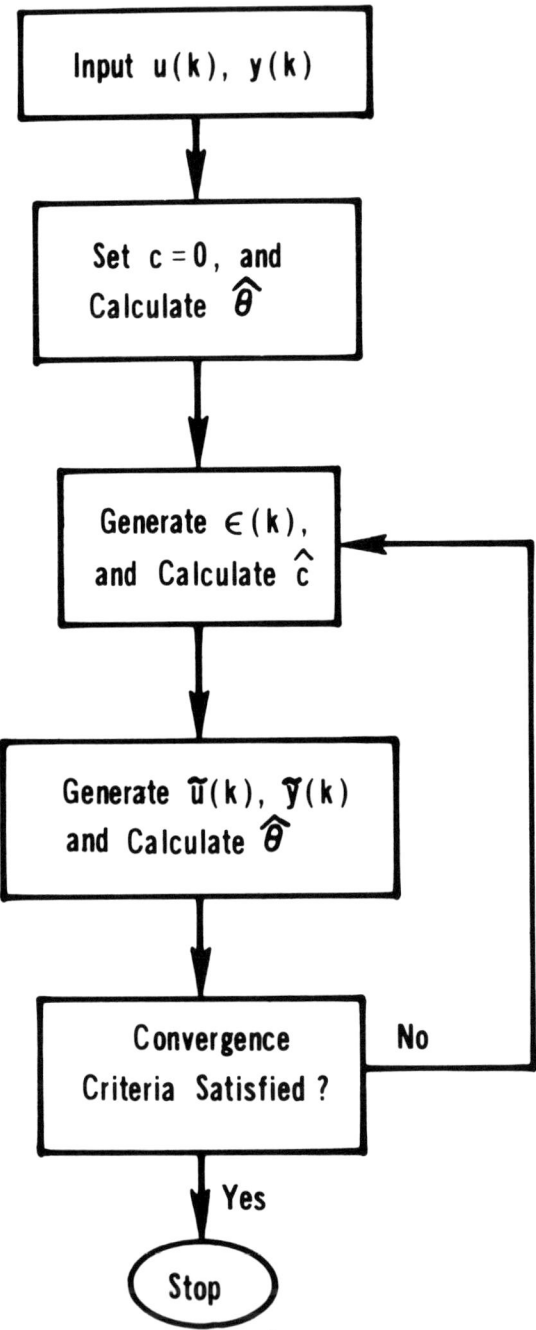

Figure 6-4. Flow Diagram of the GLS Identification Algorithm

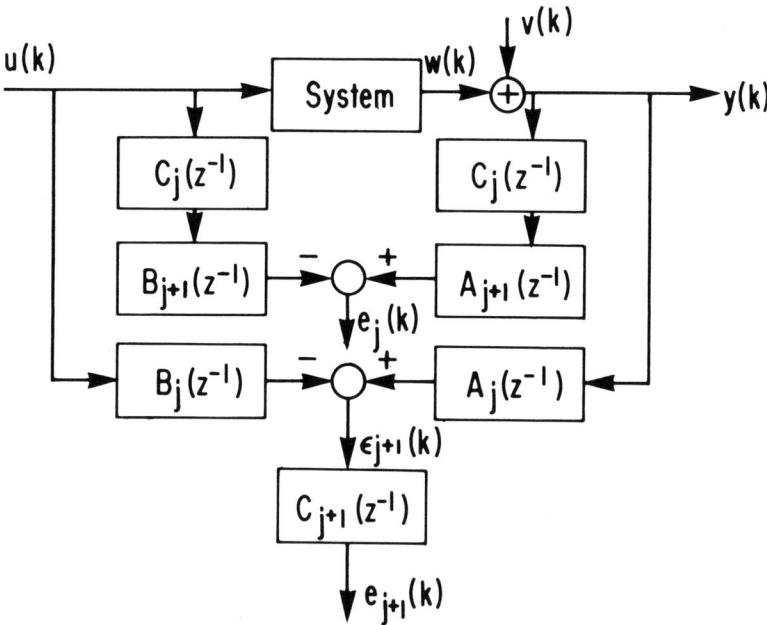

Figure 6-5. Block Diagram Interpretation of the GLS Algorithm

We note that the GLS algorithm not only identifies the system transfer function, but also provides the noise transfer function estimate $1/\hat{A}(z^{-1})\hat{C}(z^{-1})$.

Example

We will now present a simple identification problem to illustrate the GLS algorithm. Consider a second-order system with the following parameter polynomials:

$$A(q^{-1}) = 1.0 - 0.5q^{-1} + 0.5q^{-2} \quad B(q^{-1}) = 1.0$$

Thus $\theta = [-0.5, 0.5, 1.0]^T$. The system input $\{u(k)\}$ is a sequence of independent Gaussian random variables with unity variance and zero mean value. The output noise $v(k)$ is generated from the equation $C(q^{-1})A(q^{-1})v(k) = e(k)$ with $C(q^{-1}) = 1.0 + 0.85q^{-1}$. Hence $c = c_1 = 0.85$. The large c_1 value is chosen to make the residual highly correlated. $e(k)$ is a white zero-mean Gaussian random process with variance $\sigma^2 = 0.64$. This yields a low signal-to-noise ratio of $\sigma_w^2/\sigma_v^2 = 1.423$. A record of $N = 300$ data pairs $\{u(k), y(k)\}$ is used for identification.

The GLS algorithm is applied to a model having exactly the same structure as the system. During each iteration, the mean square error $\mathbf{e}^T\mathbf{e}/N$ is computed to indicate the convergence. This result is plotted in figure 6–6. It is shown that the algorithm is converged after five iterations, and also that the LS estimate $\hat{\boldsymbol{\theta}}_{LS}$ is highly biased as indicated by the correspondingly large mean-square-error value.

6.6 Remarks

1. As we have pointed out earlier, the GLS algorithm is an iterative scheme for solving the highly nonlinear minimization problem defined by equation (6.17). Therefore, the convergence of the algorithm to the optimal solution is not always guaranteed. The typical problem of this kind is that more than one local minimum in J may exist. Thus the choice of an initial condition of the parameter estimates near the optimal solution is crucial to success in finding the global minimum. In the absence of a priori information, the LS estimates are regarded as good initial conditions.

It has been shown that unimodality of J is strongly influenced by the noise level in the system (Söderström 1974a). When the signal-to-noise ratio is sufficiently high, J has a unique minimum so that the algorithm always converges to the true parameter values. But for low signal-to-noise ratio cases, J becomes nonunimodal. The GLS result depends on the initial parameter estimates used. We need to pick among all local minimum solutions the set of parameter values that provide the smallest fitting error between model and data.

The rate of convergence of the GLS algorithm is regarded as slow. This is clearly exhibited in figure 6–6. Therefore, large numbers of iterations are needed if great precision of parameter estimates is called for. However, alternative approaches are available that are more computationally efficient. These methods are introduced in the next chapter.

2. It is possible to derive an on-line version for the off-line GLS algorithm (Hasting-James and Sage 1969). This on-line estimation algorithm consists of a two-part recursive LS routine for updating $\hat{\boldsymbol{\theta}}(N)$ and $\hat{\mathbf{c}}(N)$ as N increases. Due to the fact that $\hat{\boldsymbol{\theta}}(N)$ and $\hat{\mathbf{c}}(N)$ are time-varying during each updating pass, the filtered signals $\tilde{u}(k)$ $\tilde{y}(k)$ and the residual $\epsilon(k)$ are then essentially being generated by time-varying systems. Thus the real-time LS algorithms must be used for updating $\hat{\boldsymbol{\theta}}(N)$ and $\hat{\mathbf{c}}(N)$, yielding the following on-line algorithm (see chapter 5):

$$\hat{\boldsymbol{\theta}}(N + 1) = \hat{\boldsymbol{\theta}}(N) + \gamma(N + 1)\bar{\mathbf{p}}(N)\tilde{\mathbf{x}}(N + 1)$$
$$\cdot [\tilde{y}(N + 1) - \tilde{\mathbf{x}}^T(N + 1)\hat{\boldsymbol{\theta}}(N)]$$

$$\tilde{\mathbf{c}}(N + 1) = \tilde{\mathbf{c}}(N) + \beta(N + 1)\Lambda(N)\boldsymbol{\xi}(N + 1)$$
$$\cdot [\epsilon(N + 1) - \boldsymbol{\xi}^T(N + 1)\tilde{\mathbf{c}}(N)]$$

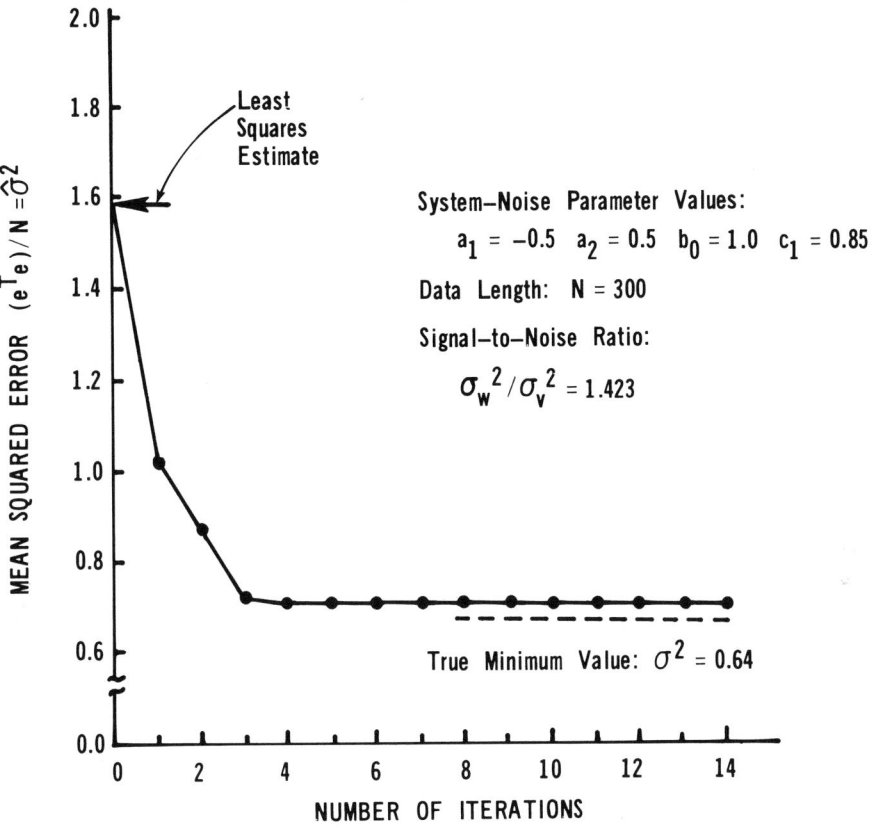

Figure 6-6. Rate of Convergence of GLS Algorithm for One Typical Computer Run of the Example System

where

$$\tilde{y}(N+1) = y(N+1) + \sum_{i=1}^{p} \tilde{c}_i(N) y(N+1-i)$$

$$\tilde{u}(N+1) = u(N+1) + \sum_{i=1}^{p} \tilde{c}_i(N) u(N+1-i)$$

$$\tilde{x}(N+1) = [-\tilde{y}(N), \ldots, -\tilde{y}(N+1-n), \tilde{u}(N+1), \\ \ldots, \tilde{u}(N+1-n)]^T$$

$$\tilde{P}(N+1) = \frac{1}{\lambda}[\tilde{P}(N) - \gamma(N+1)\tilde{P}(N)\tilde{x}(N+1) \\ \cdot \tilde{x}^T(N+1)\tilde{P}(N)], \; 0 < \lambda < 1$$

$$\gamma(N + 1) = 1/[1 + \tilde{x}^T(N + 1)\tilde{P}(N)\tilde{x}(N + 1)]$$

$$\epsilon(N + 1) = y(N + 1) + \sum_{i=1}^{n} \hat{a}_i(N + 1)y(N + 1 - i)$$

$$- \sum_{i=0}^{n} \hat{b}_i(N + 1)u(N + 1 - i)$$

$$\xi(N + 1) = [-\epsilon(N), -\epsilon(N - 1), \ldots, -\epsilon(N - p + 1)]^T$$

$$\Lambda(N + 1) = \frac{1}{\lambda}[\Lambda(N) - \beta(N + 1)\Lambda(N)\xi(N + 1)\xi^T(N + 1)\Lambda(N)]$$

$$\beta(N + 1) = 1/[1 + \xi^T(N + 1)\Lambda(N)\xi(N + 1)]$$

The corresponding error function is

$$J = \sum_{i=1}^{N} \lambda^{N-i} e^2(i)$$

3. We show below that the GLS estimates are equivalent to the Markov estimates discussed in chapter 3. This relationship further establishes the GLS estimate properties of consistency and efficiency. The Markov estimates are obtained by minimizing the weighted function J_W

$$J_W = \epsilon^T W \epsilon$$

where the weighted matrix is equal to the inverse of the covariance matrix **R** of the residual vector ϵ, or

$$W^{-1} = E[\epsilon \epsilon^T] = R$$

By equation (6.12), we can relate ϵ and **e** by $F\epsilon = e$ where **F** has the following form:

$$F = \begin{bmatrix} 1 & & & & & & \\ c_1 & 1 & & & & & \\ c_2 & c_1 & 1 & & & 0 & \\ \vdots & & & & & & \\ c_p & \ldots & c_1 & 1 & & & \\ \vdots & & & & & & \\ 0 & \ldots & & & c_p & \ldots & 1 \end{bmatrix}$$

Similarly, we can show that
$$\tilde{y} = Fy \quad \tilde{X} = FX$$
F is a nonsingular square matrix. Let $F^{-1} = H$; then $\epsilon = He$. Thus
$$R = E[\epsilon\epsilon^T] = E[Hee^T H]$$
Because $E[ee^T] = \sigma^2 I$, we have $R = \sigma^2 HH^T$. Consequently $R^{-1} = (1/\sigma^2)F^T F$. Thus

$$J_W = \epsilon^T W\epsilon = (y - X\theta)^T R^{-1}(y - X\theta)$$
$$= \frac{1}{\sigma^2}(y - X\theta)^T F^T F(y - X\theta)$$
$$= \frac{1}{\sigma^2}(Fy - FX\theta)^T(Fy - FX\theta)$$
$$= \frac{1}{\sigma^2}(\tilde{y} - \tilde{X}\theta)^T(\tilde{y} - \tilde{X}\theta)$$
$$= \frac{1}{\sigma^2}e^T e$$

which is equivalent to the GLS error function. Hence GLS estimates are the same as Markov estimates.

4. One commonly encountered case in system identification is that the output additive disturbance $v(k)$ is a white noise (Steiglitz and McBride 1965). In this case, we can denote $v(k) = e(k)$, and the system equation becomes

$$A(q^{-1})y(k) = B(q^{-1})u(k) + A(q^{-1})e(k)$$

Now the whitening filter transfer function is $1/A(z^{-1})$ as shown in figure 6–7a. At the $(j + 1)$th iteration of GLS algorithm, we have

$$\frac{A_{j+1}(q^{-1})}{\hat{A}_j(q^{-1})} y(k) = \frac{B_{j+1}(q^{-1})}{\hat{A}_j(q^{-1})} u(k) + e(k)$$

This yields the following equation at convergence:

$$y(k) = \frac{\hat{B}(q^{-1})}{\hat{A}(q^{-1})} u(k) + e(k)$$

which is illustrated in figure 6–7b.

One interesting aspect of the above result is that the generalized equation error $e(k)$ can also be viewed as the output error between the system and the model. Thus we have achieved minimum-output-error identifica-

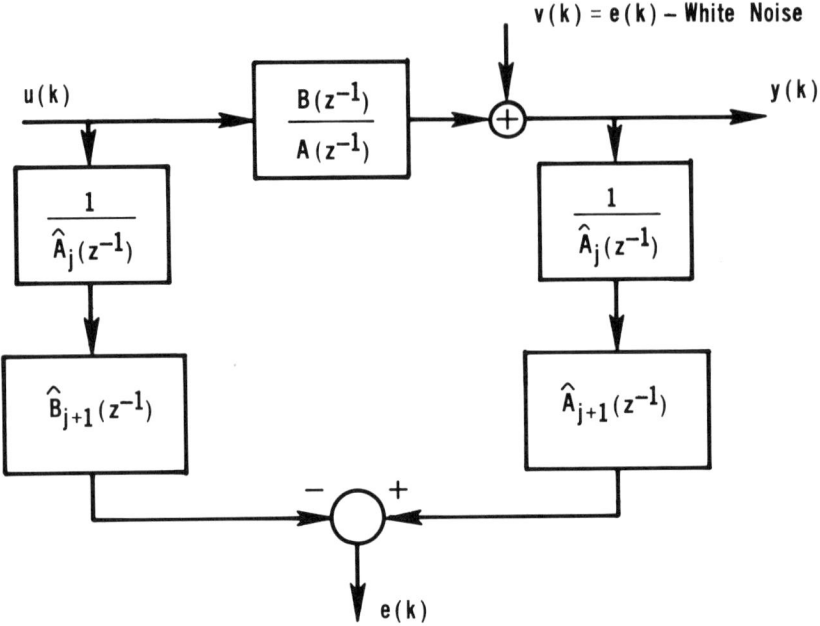

Figure 6-7a. GLS Algorithm for the Case in Which $v(k)$ is White Noise

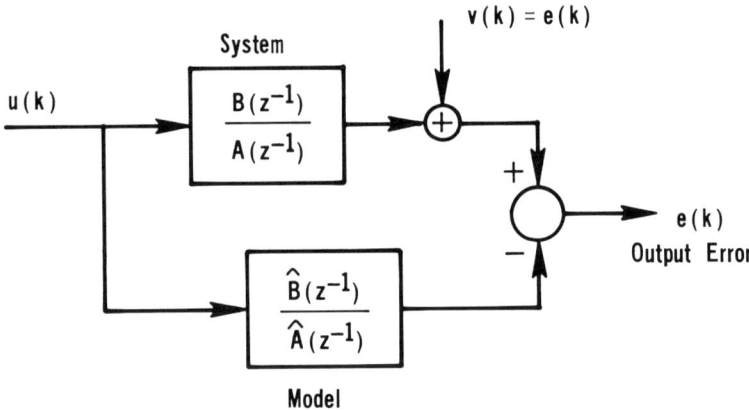

Figure 6-7b. Minimum-output-error Identification Interpretation of GLS at Convergence

tion by way of the GLS algorithm. Minimum-output-error identification has a very strong intuitive appeal.

5. The GLS method leads to a noise transfer function $1/A(z^{-1})C(z^{-1})$ as shown in figure 6-2. It is possible to obtain different forms of the transfer function by assuming different models for the residual $\epsilon(k)$—see, for ex-

ample, equation (6.28). A popular form for the noise transfer function is shown in figure 7-3. The estimation procedures for that model are discussed in section 7.2.

6. There are a number of approaches that can be used to solve the nonlinear minimization problem associated with the generalized error function J. The GLS algorithm introduced above is the one most widely used. One other closely related GLS algorithm (Clarke 1967) is illustrated in figure 6-8. This algorithm differs in that a chain of whitening filters are used instead of a single filter as in figure 6-5. Therefore, at the jth iteration, the model equation is

$$\hat{A}_j(q^{-1})y(k) = \hat{B}_j(q^{-1})u(k) + \frac{1}{\prod_{i=1}^{j}\hat{c}_i(q^{-1})}\epsilon_j(k)$$

in which each $\hat{c}_i(q^{-1})$ is a pth-order polynomial. $\hat{A}_j(q^{-1})$, $\hat{B}_j(q^{-1})$, and $\hat{c}_j(q^{-1})$ are estimated as before. At convergence,

$$\lim_{j\to\infty} \hat{c}_j(q^{-1}) = 1$$

which means that $\epsilon_\infty(k)$ is white and the input-output signals are no longer filtered. Again, analysis has shown that this algorithm does not always converge at the correct answer (Söderström 1974b).

6.7 An Alternative GLS Solution Technique

The GLS algorithms introduced above are characterized by repeated filtering of the system's input-output signals. In this section, we present an alternative technique to solve the GLS problem that does not require repeated data filtering (Hsia 1976). This means that it is computationally more efficient. The technique takes the approach of directly removing the bias in the LS estimates. Furthermore, the new approach leads to a simpler recursive algorithm for on-line system identification.

From equations (6.4) and (6.13) we get

$$\mathbf{y} = \mathbf{X}\boldsymbol{\theta} + \boldsymbol{\epsilon}$$

$$\boldsymbol{\epsilon} = \boldsymbol{\Omega}\mathbf{c} + \mathbf{e}$$

Combining the above two equations yields

$$\mathbf{y} = [\mathbf{X}, \boldsymbol{\Omega}]\begin{bmatrix}\boldsymbol{\theta}\\\mathbf{c}\end{bmatrix} + \mathbf{e} \qquad (6.20a)$$

The parameter estimates that minimizes J

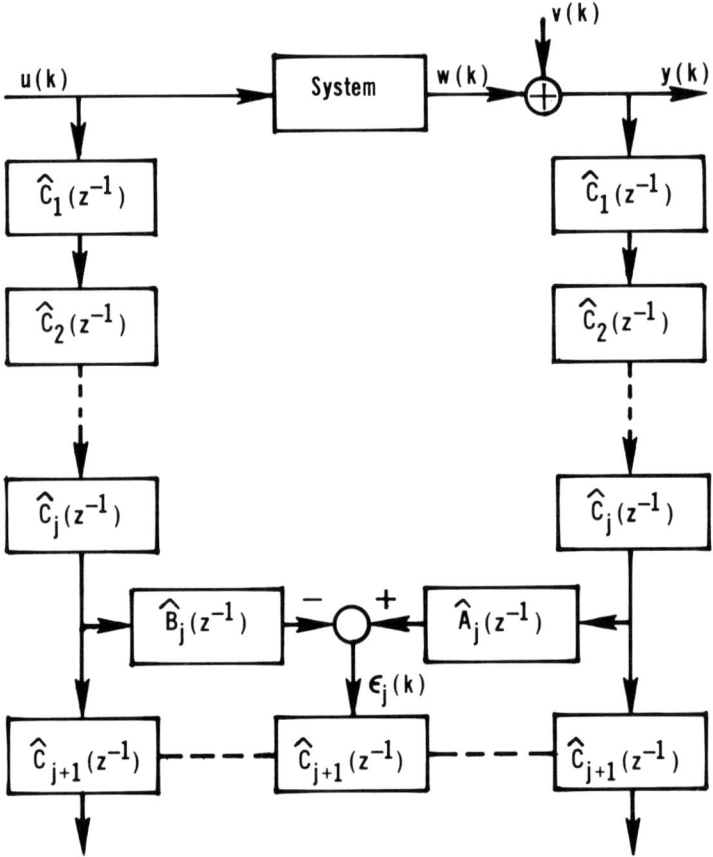

Figure 6–8. A Different Version of the GLS Algorithm

$$J = e^T e \tag{6.20b}$$

can be expressed as

$$\begin{bmatrix} \hat{\theta} \\ \hat{c} \end{bmatrix} = \begin{bmatrix} X^T X & X^T \Omega \\ \Omega^T X & \Omega^T \Omega \end{bmatrix}^{-1} \begin{bmatrix} X^T y \\ \Omega^T y \end{bmatrix} \tag{6.21}$$

Now apply the partitioned matrix inverse identity (see section 3.6) to the above to get

$$\hat{\theta} = (X^T X)^{-1} X^T y - (X^T X)^{-1} X^T \Omega D^{-1} \Omega^T M y$$

$$\hat{c} = D^{-1} \Omega^T M y \tag{6.22}$$

where
$$M = I - X(X^TX)^{-1}X^T$$
$$D = \Omega^T\Omega - \Omega^TX(X^TX)^{-1}X^T\Omega$$
$$= \Omega^TM\Omega$$

$\hat{\boldsymbol{\theta}}$ in equation (6.22) can also be written as
$$\hat{\boldsymbol{\theta}} = (X^TX)^{-1}X^Ty - (X^TX)^{-1}X^T\Omega\hat{c}$$

We recognize that the first term in $\hat{\boldsymbol{\theta}}$ is the LS estimate of $\boldsymbol{\theta}$, and the second term is the bias. Thus

$$\hat{\boldsymbol{\theta}} = \hat{\boldsymbol{\theta}}_{LS} - \hat{\boldsymbol{\theta}}_{BIAS} \qquad (6.23)$$

These results show that a consistent estimate $\hat{\boldsymbol{\theta}}$ can be obtained if we subtract the bias from the LS estimate. However, the bias $\hat{\boldsymbol{\theta}}_{BIAS}$ has to be estimated because the residual matrix Ω is unknown. This leads to the following iterative estimation algorithm.

Step 1. Assume $\hat{c} = 0$ initially, compute $\hat{\boldsymbol{\theta}}_{LS}$ from

$$\hat{\boldsymbol{\theta}}_{LS} = (X^TX)^{-1}X^Ty$$

Set $\hat{\boldsymbol{\theta}} = \hat{\boldsymbol{\theta}}_{LS}$. Note that the quantity $\Gamma = (X^TX)^{-1}X^T$ is invariant throughout the algorithm, so we only compute it once.

Step 2. Generate the residuals $\epsilon(k)$ by

$$\epsilon = y - X\hat{\boldsymbol{\theta}}$$

Then define Ω and compute $D = \Omega^TM\Omega$

Step 3. Now calculate $\hat{c}$ and $\hat{\boldsymbol{\theta}}_{BIAS}$ from

$$\hat{c} = D^{-1}\Omega My$$
$$\hat{\boldsymbol{\theta}}_{BIAS} = \Gamma\Omega\hat{c}$$

Then update $\hat{\boldsymbol{\theta}}$ by

$$\hat{\boldsymbol{\theta}} = \hat{\boldsymbol{\theta}}_{LS} - \hat{\boldsymbol{\theta}}_{BIAS}$$

Step 4. Go back to step 2 and repeat the computations.

We see that the above iterative algorithm is essentially an operation to successively improve the accuracy of the bias term that corrects the LS parameter estimate. This algorithm can be easily extended to identify multivariable systems (Hsia 1976), whereas the GLS algorithms introduced earlier would fail because of the data filtering requirements.

The main appeal of this algorithm is its computational simplicity relative to the GLS algorithms. We observe that in the GLS algorithms there are two matrix inversions plus data filtering in each iteration, whereas in this algorithm only one matrix inversion is called for per iteration, and no data filtering is needed. However, these computational advantages are somewhat offset by the excessive computations required by the matrices **M** and **D**. In order to avoid **M** and **D**, we use the following approximate solution for $\hat{c}$:

$$\hat{c} = (\Omega^T \Omega)^{-1} \Omega^T \epsilon \qquad (6.24)$$

which is far easier to compute. The justification of equation (6.24) is as follows.

If we use the approximation $\theta = \hat{\theta}$ in equation (6.20a), we get

$$y - X\hat{\theta} = \Omega c + e$$

This yields $\epsilon = \Omega c + e$, which leads to the LS estimate $\hat{c}$ in equation (6.24). This result is the same as that in the GLS algorithm. Thus we have the modified algorithm given by

$$\hat{\theta} = \hat{\theta}_{LS} - \Gamma \Omega \hat{c}$$
$$\hat{c} = (\Omega^T \Omega)^{-1} \Omega^T \epsilon \qquad (6.25)$$

Example

We now present a simple example to demonstrate the performance of these bias-correction algorithms. In particular, we wish to compare them with the GLS algorithm in terms of computational efficiency and rate of convergence. Thus we choose to study the same system investigated in section 6.5, that is, the system with the following parameter polynomials:

$$A(q^{-1}) = 1.0 - 0.5q^{-1} + 0.5q^{-2}, \quad B(q^{-1}) = 1.0$$
$$C(q^{-1}) = 1.0 + 0.85q^{-1}$$

Figure 6–9 shows the results of all three algorithms under the same experimental conditions as described in the earlier example. We see that the characteristic initial peaking of the bias-correction algorithm slows down the convergence rate considerably. The modified algorithm, on the other hand, has convergence that is much improved, but still slower than the GLS algorithm. All three algorithms converged to almost the same solution at the sixth iteration. Table 6–1 gives the computation time and memory requirements of the two bias-correction algorithms measured against

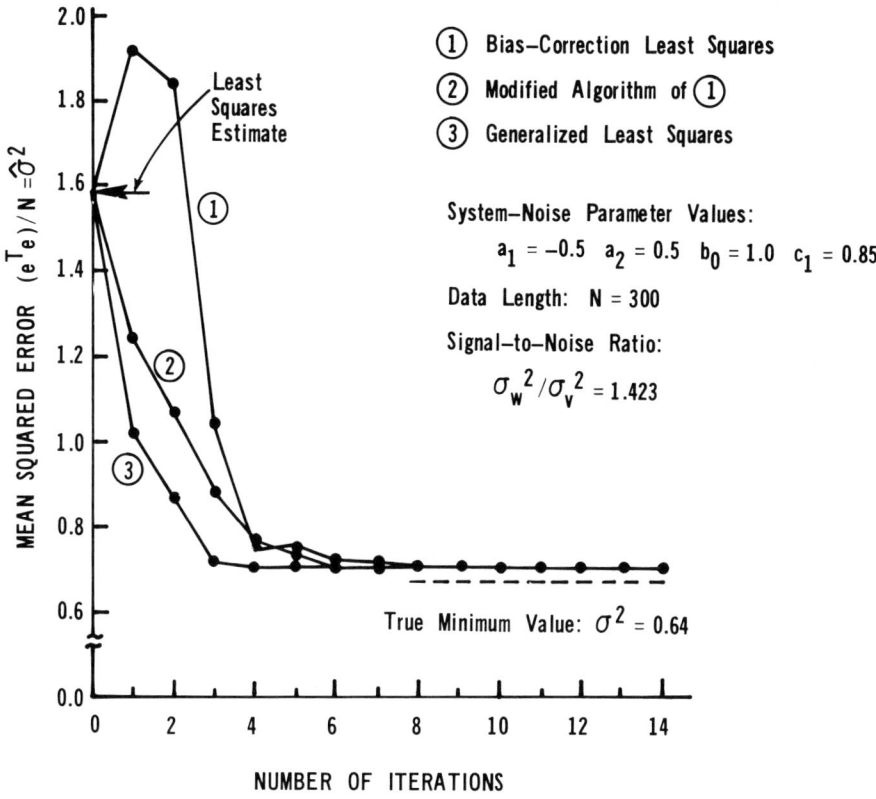

Figure 6-9. Rate of Convergence of Three Identification Algorithms for One Typical Computer Run of the Example System

Table 6-1
Computation Time and Memory Requirements of the Bias-correction Algorithms Relative to Those of the GLS Algorithm

	Bias-correction least-squares algorithm		*Modified algorithm*	
Data length N	300	500	300	500
Processor time per iteration	75%	67%	37%	33%
Virtual memory	99%	99%	90%	90%
	GLS algorithm is 100%			

Note: Based on requirements of 20 iterations on the Burroughs 6700 Computer.

the same requirements of the GLS algorithm. The results show that the bias-correction algorithms are indeed more efficient than the GLS algorithm.

The above algorithms are off-line solutions to the estimation problem associated with equation (6.20). It is possible, however, to derive a recursive solution for on-line identification. This solution calls for simultaneous estimation of both θ and c. Thus we redefine equation (6.20a) as $y = \Phi\beta + \epsilon$ where

$$\Phi = [X \vdots \Omega], \quad \beta = \begin{bmatrix} \theta \\ \cdots \\ c \end{bmatrix}$$

This yields the LS solution of β as

$$\hat{\beta} = (\Phi^T\Phi)^{-1}\Phi^{-1}y$$

The above solution has a known recursive form:

$$\hat{\beta}(N+1) = \hat{\beta}(N) + \gamma(N+1)P(N)\phi(N+1)$$
$$\cdot [y(N+1) - \phi^T(N+1)\hat{\beta}(N)]$$
$$P(N+1) = P(N) - \gamma(N+1)P(N)\phi(N+1)\phi^T(N+1)P(N)$$
$$\gamma(N+1) = 1/[1 + \phi^T(N+1)P(N)\phi^T(N+1)]$$

(6.26)

where

$$\phi(N+1) = [-y(N), \ldots,$$
$$-y(N+1-n), u(N+1), \ldots u(N+1-n),$$
$$-\epsilon(N), \ldots, -\epsilon(N+1-p)]$$

$$\epsilon(N+1) = y(N+1) + \sum_{i=1}^{n} \hat{a}_i(N+1)y(N+1-i)$$
$$- \sum_{i=0}^{n} \hat{b}_i(N+1)u(N+1-i)$$

This on-line algorithm is considerably simpler than that of GLS given in section 6.6. As an example, the algorithm is applied to the second-order system studied above, and the identification result is plotted in figure 6–10. In this plot, the normalized parameter error is computed for each iteration as an indication of convergence.

Finally, we wish to note that the above on-line system-noise parameter identification algorithm can be generalized in two directions (Talmon and Van den Boom 1973):

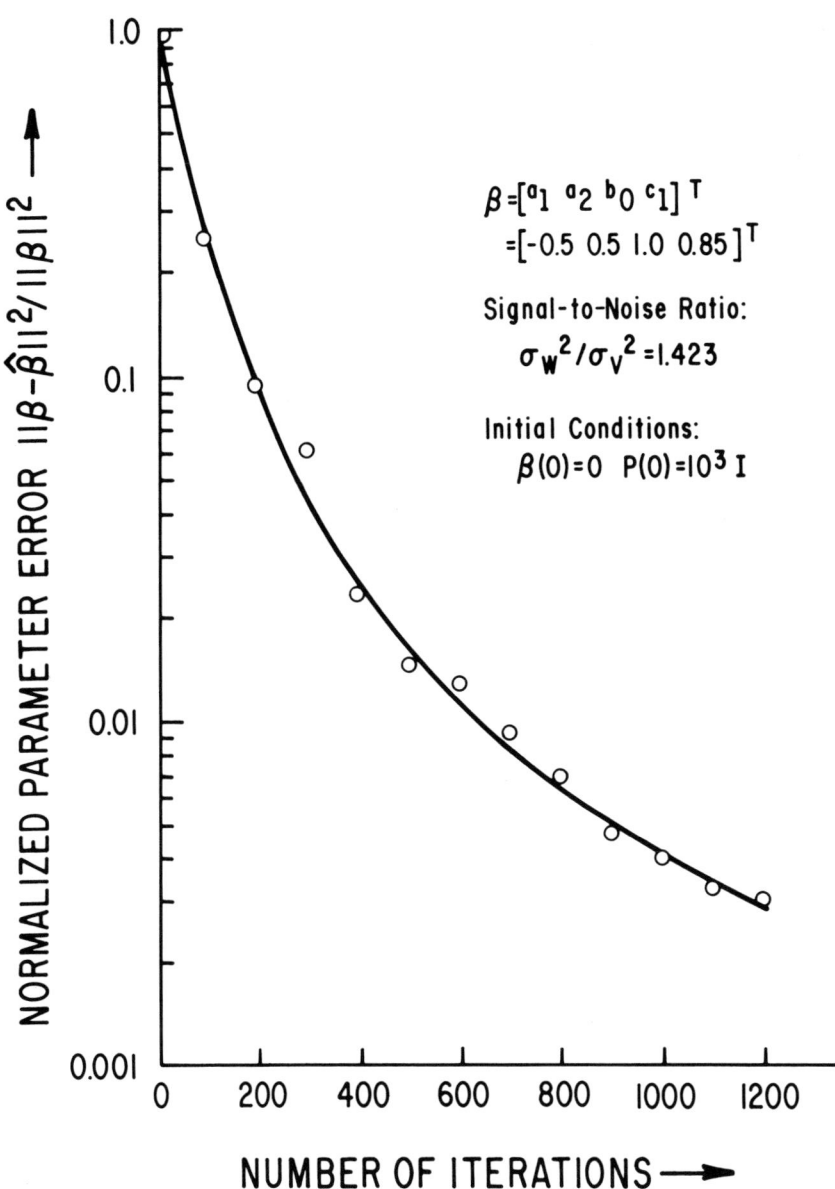

Figure 6–10. Performance of the System-noise On-line Identification Algorithm

1. Assume a mixed autoregressive moving-average model for the residual $\epsilon(k)$, that is

$$\epsilon(k) = \frac{D(q^{-1})}{C(q^{-1})} e(k) \tag{6.27}$$

where

$$D(q^{-1}) = 1 + d_1 q^{-1} + \ldots + d_s q^{-s}.$$

We see that both the autoregressive model [where $D(q^{-1}) = 1$] and the moving-average model [where $C(q^{-1}) = 1$] are special cases. The advantage of using the general noise model is that the total number of noise parameters needed in modeling $\epsilon(k)$ is always less than that needed in either of the two special models. This helps to reduce the number of parameters to be estimated, and hence simplifies the algorithm.

2. Apply the exponentially weighted error function

$$J = \sum_{i=1}^{N} \lambda^{N-i} e^2(i) \quad 0 < \lambda < 1 \tag{6.28}$$

This error function is more appropriate for the recursive estimation because the parameter estimates are changing from iteration to iteration, and the latest estimates (which are more accurate) should be given the most weight. This strategy generally has the effect of improving the stability and convergence rate of the recursive algorithm.

Based on equations (6.27) and (6.28), we can derive the following generalized recursive estimation algorithm (Talmon and Van den Boom 1973):

$$\hat{\boldsymbol{\delta}}(N+1) = \hat{\boldsymbol{\delta}}(N) + \gamma(N+1)\mathbf{P}(N)\boldsymbol{\phi}(N+1)$$
$$\cdot [y(N+1) - \boldsymbol{\phi}^T(N+1)\hat{\boldsymbol{\delta}}(N)]$$

$$\mathbf{P}(N+1) = \frac{1}{\lambda}[\mathbf{P}(N) - \gamma(N+1)\mathbf{P}(N)\boldsymbol{\phi}(N+1)\boldsymbol{\phi}^T(N+1)\mathbf{P}(N)]$$

$$\gamma(N+1) = 1/[1 + \boldsymbol{\phi}^T(N+1)\mathbf{P}(N)\boldsymbol{\phi}(N+1)]$$

where

$$\hat{\boldsymbol{\delta}}^T = [\hat{a}_1, \ldots, \hat{a}_n, \hat{b}_0, \ldots, \hat{b}_n, \hat{c}_1, \ldots, \hat{c}_p, \hat{d}_1, \ldots, \hat{d}_s]$$

$$\boldsymbol{\phi}^T(N+1) = [-y(N), \ldots,$$
$$-y(N+1-n), u(N+1), \ldots, u(N+1-n),$$
$$-\epsilon(N), \ldots, -\epsilon(N+1-p), e(N), \ldots, e(N+1-s)]$$

$$\epsilon(N+1) = y(N+1) + \sum_{i=1}^{n} \hat{a}_i(N+1)y(N+1-i)$$

$$- \sum_{i=0}^{n} \hat{b}_i(N+1)u(N+1-i)$$

$$e(N+1) = \epsilon(N+1) + \sum_{i=1}^{p} \hat{c}_i(N+1)\epsilon(N+1-i)$$

$$- \sum_{i=1}^{s} \hat{d}_i(N+1)e(N+1-i)$$

Computer-simulated experimental results using the above algorithm can be found in the literature (Talmon and Van den Boom 1973). There is no theoretical proof for the convergence of these algorithms.

6.8 Instrumental Variable Method

We have shown in the previous sections that the GLS approach is useful in removing bias in the parameter estimates when the residual in the system equation is autocorrelated. However, the iterative GLS algorithms are considerably more complicated than the simple one-shot LS solution. The instrumental variable (IV) method to be discussed below is a method that is as simple as the LS method, and yet it yields consistent estimates (Wong and Polâk 1967). Therefore, it is of sufficient interest that we give a brief review of the method.

Consider again the system equation

$$\mathbf{y} = \mathbf{X}\boldsymbol{\theta} + \boldsymbol{\epsilon} \qquad (6.29)$$

Suppose there exists a $(2n+1) \times N$ matrix $\mathbf{Z}$ (the same dimension as $\mathbf{X}$) which satisfies the following limiting properties:

$$\lim_{N \to \infty} \frac{1}{N} \mathbf{Z}^T \boldsymbol{\epsilon} = \mathbf{0}$$

$$\lim_{N \to \infty} \frac{1}{N} \mathbf{Z}^T \mathbf{X} = \mathbf{Q} \qquad (6.30)$$

$\mathbf{Q}$ is nonsingular.

Premultiply equation (6.29) by $\mathbf{Z}^T$ and solve for $\boldsymbol{\theta}$

$$\boldsymbol{\theta} = (\mathbf{Z}^T\mathbf{X})^{-1}\mathbf{Z}^T\mathbf{y} - (\mathbf{Z}^T\mathbf{X})^{-1}\mathbf{Z}^T\boldsymbol{\epsilon} \qquad (6.31)$$

Then we take

$$\hat{\boldsymbol{\theta}}_{IV} = (\mathbf{Z}^T\mathbf{X})^{-1}\mathbf{Z}^T\mathbf{y} \qquad (6.32)$$

as the estimate of $\boldsymbol{\theta}$. We call this estimate $\hat{\boldsymbol{\theta}}_{IV}$ the *instrumental variable estimate*, the matrix $\mathbf{Z}$ the *instrumental variable matrix*, and the elements in $\mathbf{Z}$ the *instrumental variables*.

We clearly see that equation (6.32) is of the same form as the LS solution. Thus it is simple to compute. We will now show that $\hat{\boldsymbol{\theta}}_{IV}$ is a consistent estimate of $\boldsymbol{\theta}$, that is, that the following limit exists:

$$\lim_{N\to\infty} \hat{\boldsymbol{\theta}}_{IV} = \boldsymbol{\theta}$$

From equation (6.31), we write

$$\hat{\boldsymbol{\theta}}_{IV} = \boldsymbol{\theta} + (\mathbf{Z}^T\mathbf{X})^{-1}\mathbf{Z}^T\boldsymbol{\epsilon}$$

Take the limit $N \to \infty$ on both sides:

$$\lim_{N\to\infty} \hat{\boldsymbol{\theta}}_{IV} = \boldsymbol{\theta} + \lim_{N\to\infty}\left(\frac{\mathbf{Z}^T\mathbf{X}}{N}\right)^{-1} \lim_{N\to\infty}\left(\frac{\mathbf{Z}^T\boldsymbol{\epsilon}}{N}\right)$$

Invoke the properties in equation (6.30), and we get the desired result

$$\lim_{N\to\infty} \hat{\boldsymbol{\theta}}_{IV} = \boldsymbol{\theta}$$

The remaining important question is how to determine the instrumental variables needed to construct $\mathbf{Z}$. The properties of $\mathbf{Z}$ given in equation (6.30) can be simply interpreted to mean that the instrumental variables are uncorrelated with $\boldsymbol{\epsilon}$ (or with the random disturbances $\mathbf{v}$), but strongly correlated with $u(k)$ and $y(k)$ in $\mathbf{X}$. One obvious ideal choice of $\mathbf{Z}$ then is—assuming the input $u(k)$ is persistently exciting of order $2n + 1$—

$$\mathbf{Z} = \begin{bmatrix} -w(n), & \ldots, & -w(1) & \vdots & u(n+1), & \ldots, & u(1) \\ -w(n+1), & \ldots, & -w(2) & \vdots & u(n+2), & \ldots, & u(2) \\ \vdots & & & & \vdots & & \\ -w(n+N-1), & \ldots, & -w(N) & \vdots & u(n+N), & \ldots, & u(N) \end{bmatrix} \qquad (6.33)$$

where $w(k)$ is the exact system output (see figure 6–11). In fact, it can be shown that this $\mathbf{Z}$ is the best choice in the sense of minimax estimation error for $\hat{\boldsymbol{\theta}}_{IV}$ (Wong and Polak 1967). However, the system output sequence $\{w(k)\}$ is actually not accessible, so they must be approximated.

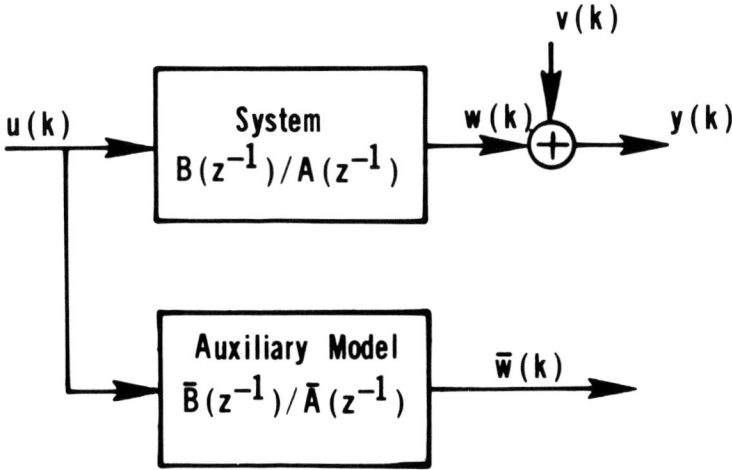

Figure 6-11. Generation of Instrumental Variables Using an Auxiliary Model

One simple approximation is to use the output $\{\bar{w}(k)\}$ of any auxiliary model whose transfer function $\bar{B}(z^{-1})/\bar{A}(z^{-1})$ is an estimate of the true system transfer function $B(z^{-1})/A(z^{-1})$. In fact it can be shown (Finigan and Rowe 1974) that any auxiliary model $\bar{B}(z^{-1})/\bar{A}(z^{-1})$ that is stable and without common factors will be acceptable.

It is obvious that there can be many ways for selecting instrumental variables that satisfy the general conditions given in equation (6.30). We can expect that a different choice of $\mathbf{Z}$ leads to different efficiency for the estimate $\hat{\boldsymbol{\theta}}_{IV}$. But in any case, the efficiency of $\hat{\boldsymbol{\theta}}_{IV}$ is always inferior than that of $\hat{\boldsymbol{\theta}}_{GLS}$. The advantage of the IV method is that it is less complicated to use.

Finally, we wish to present the on-line version of the IV solution given in equation (6.32). Because of their similarity to the LS solution, the following recursive algorithms can be easily derived (Wong and Polak 1967). We assume that $\mathbf{Z}$ is constructed as in equation (6.34) with $\bar{w}(k)$ replacing $w(k)$:

$$\hat{\boldsymbol{\theta}}_{IV}(N + 1) = \hat{\boldsymbol{\theta}}_{IV}(N) + \gamma(N + 1)\mathbf{P}(N)\bar{\mathbf{z}}(N + 1)$$
$$\cdot [y(N + 1) - \mathbf{x}^T(N + 1)\hat{\boldsymbol{\theta}}_{IV}(N)]$$

$$\mathbf{P}(N + 1) = \mathbf{P}(N) - \gamma(N + 1)\mathbf{P}(N)\bar{\mathbf{z}}(N + 1)\mathbf{x}^T(N + 1)\mathbf{P}(N)$$

$$\gamma(N + 1) = 1/[1 + \mathbf{x}^T(N + 1)\mathbf{P}(N)\bar{\mathbf{z}}(N + 1)]$$

where

$$\boldsymbol{\theta}_{IV} = [a_1, \ldots a_n, b_0, \ldots, b_n]^T$$

$$\mathbf{x}(N+1) = [-y(N), \ldots,$$
$$-y(N+1-n), u(N+1), \ldots, u(N+1-n)]^T$$

$$\bar{\mathbf{z}}(N+1) = [-\bar{w}(N), \ldots,$$
$$-\bar{w}(N+1-n), u(N+1), \ldots, u(N+1-n)]^T$$

Various ways of generating the sequence $\{\bar{w}(k)\}$ can be found in the literature (Wong and Polak 1967; Young 1970; Pandya 1974).

6.9 Concluding Remarks

The main purpose of this chapter was to show that LS estimations are asymptotically biased when the residuals are autocorrelated. This condition generally exists in practice. To deal with this bias problem, GLS methods were introduced in which an appropriate noise model was incorporated in the estimation process. The GLS estimates were shown to be consistent.

In addition, the IV method was discussed as an alternative approach to solving the correlated residual problem. We note that there exist many other approaches in the literature that are also very effective for parameter identification. An important one among them is the maximum likelihood (ML) method (Åström and Bohlin 1965; Gertler and Bányász), in which the random error $e(k)$ is assumed to be white Gaussian. Under the same assumption, the GLS method can also be interpreted as the ML method (Söderström 1974a).

In the next chapter, we present a group of identification methods called *multistage least squares*. These methods can achieve LS identification without the iterative procedures required by the GLS method. Thus they are computationally more advantageous.

References

Åström, K.J., and Bohlin, T., "Numerical Identification of Linear Dynamic Systems from Normal Operating Records," IFAC Symposium—Theory of Self-adaptive Control Systems, Teddington, England, 1965.

Clarke, D. W., "Generalised-least-squares Estimation of Parameters of a Dynamic Model," *IFAC Symposium—Identification in Automatic Control Systems*, Papers 3, 17, Prague, 1967.

Finigan, B. M., and Rowe, I. H. "Strongly Consistent Parameter Estimation by the Introduction of Strong Instrumental Variables," *IEEE Transactions on Automatic Control*, Vol. AC-19, pp. 825–830, Dec. 1974.

Gertler, J., and Bányász, C., A Recursive (On-line) Maximum Likelihood Identification Method," *IEEE Transactions on Automatic Control*, Vol. AC-19, pp. 816–820, Dec. 1974.

Hasting-James, R., and Sage, M. W., "Recursive Generalised-least-squares Procedures for On-line Identification of Process Parameters," Proceedings IEE, Vol. 116, pp. 2057–2062, Dec. 1969.

Hsia, T. C. "On Least Squares Algorithm for System Parameter Identification," *IEEE Transactions on Automatic Control*, Vol. AC-21, pp. 104–108, Feb. 1976.

Pandya, R. N., "A Class of Bootstrap Estimators and Their Relationship to the Generalized Two Stage Least Squares Estimators," *IEEE Transactions on Automatic Control*, Vol. AC-19, pp. 831–835, Dec. 1974.

Söderström, T., "Convergence Properties of the Generalized Least Squares Identification Method," *Automatica*, Vol. 10, pp. 617–626, 1974a.

———. "On the Generalized Least Squares Method. Counter-examples to General Convergence," *Automatica*, Vol. 10, pp. 681–683, 1974b.

Steiglitz, K., and McBride, L. E., "A Technique for Identification of Linear Systems," *IEEE Transactions on Automatic Control,* Vol. Ac-10, pp. 461–464, Oct. 1965.

Talmon, J. L., and Van den Boom, A. J. W., "On the Estimation of the Transfer Function Parameters of Process-and-Noise Dynamics Using a Single-stage Estimator," 3d IFAC Symposium—Identification and System Parameter Estimation, pp. 711–720, 1973.

Wong, K. Y., and Polak, E., "Identification of Linear Discrete Time Systems Using the Instrumental Variable Approach," *IEEE Transactions on Automatic Control*, Vol. AC-12, pp. 707–718, Dec. 1967.

Young, P. C., "An Instrumental Variable Method for Real-Time Identification of Noisy Processes," *Automatica*, Vol. 6, pp. 271–287, 1970.

7

Multistage Least-squares Identification Techniques

7.1 Introduction

In chapter 6, we introduced the generalized least-squares (GLS) technique to solve the noisy system identification problem. This technique is shown to yield very accurate parameter estimates. Also briefly discussed was the instrumental variable (IV) technique, which is simpler to compute than GLS but whose estimates are less accurate. The generation of "good" instrumental variables remains an unsolved problem.

In this chapter, we introduce a different least-squares approach to deal with the correlated residual identification problem. The approach, called *multistage least squares* (MSLS), is one which the system-noise identification problem is divided into three stages, and each stage is solved by a simple least-squares procedure. The resulting estimates are consistent.

The main advantage of MSLS over GLS is the computational simplicity resulting from the absence of iteration; thus it does not encounter the convergence problem of GLS. The trade-off for computational simplicity is the somewhat reduced estimator accuracy. By comparison to the IV method, MSLS is more straightforward to apply and the estimator accuracy is more tractable.

Three MSLS techniques are introduced in this chapter, two of the three being closely related. In general, the three techniques share the same three-stage estimation structure as depicted in figure 7–1. At the end of the chapter, a comparison study is made between the MSLS and GLS techniques in terms of their estimation accuracies and computation time requirements.

7.2 MSLS Method I

Shown in figure 7–1 is an auxiliary system model being identified in the first stage of the MSLS approach. The auxiliary system model is selected in such a way that it can be consistently estimated by the LS procedure. This model then possesses the unbiased information about the system so that the system parameters can be estimated in the next LS stages. In the third LS stage, the noise parameters are estimated. The first MSLS meth-

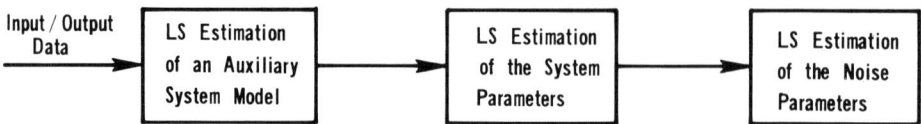

Figure 7–1. The Three-stage Estimation Scheme of the MSLS Identification Method

od, as well as method II, employs the weighting sequence as the auxiliary system model.

Consider the stable linear system shown in figure 7–2. As before, the system model is written as

$$A(q^{-1})y(k) = B(q^{-1})u(k) + \epsilon(k) \tag{7.1}$$

where

$$\epsilon(k) = A(q^{-1})v(k)$$

$$A(q^{-1}) = 1 + a_1 q^{-1} + \ldots + a_n q^{-n}$$

$$B(q^{-1}) = b_0 + b_1 q^{-1} + \ldots + b_n q^{-n}$$

where n is assumed known

It has already been shown that this model cannot be correctly identified directly by the LS method. However, as shown in chapter 4, an unbiased and consistant LS estimate of the system's weighting sequence model can be obtained whether or not the noise $v(k)$ is autocorrelated. We now present the following MSLS identification procedure (Pandya and Pagurek 1973).

Stage 1. Based on the assumption that the system is stable, the weighting sequence can be approximated by a finite number of terms. Let the weighting sequence be $\{h_k\}$, $k = 0, 1, 2, \ldots, p$, and p be sufficiently large ($p > 2n + 1$). Then, by convolution summation,

$$y(k) = \sum_{i=0}^{p} h_i u(k - i) + v(k) \tag{7.2}$$

We assume that $v(k)$ is a zero-mean random noise, either white or colored, uncorrelated to $u(k)$. Given a set of input-output data points of length $N + p$, we can write

$$\mathbf{y} = \mathbf{Uh} + \mathbf{v} \tag{7.3}$$

where

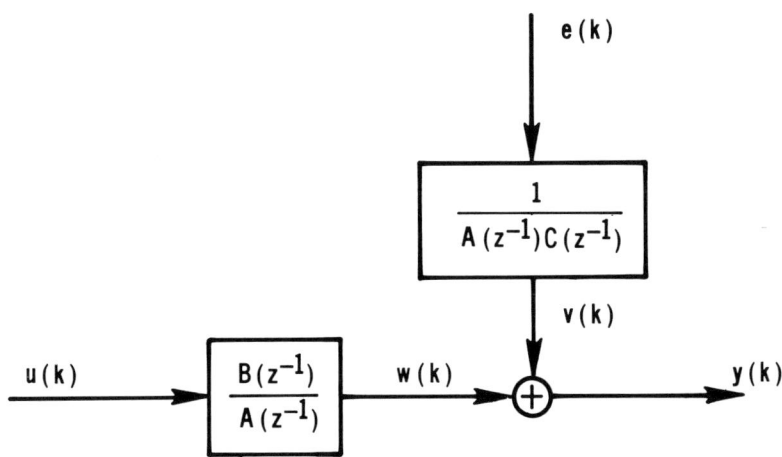

Figure 7-2. Linear System with Additive Output Noise

$$\mathbf{y} = [y(p), y(p+1), \ldots, y(p+N)]^T$$
$$\mathbf{v} = [v(p), v(p+1), \ldots, v(p+N)]^T$$
$$\mathbf{h} = [h_0, h_1, \ldots, \ldots, h_p]^T$$
$$\mathbf{U} = \begin{bmatrix} u(p), u(p-1), \ldots, u(0) \\ u(p+1), u(p), \ldots, u(1) \\ \vdots \\ u(p+N), \ldots, \ldots, u(N) \end{bmatrix}$$

From equation (7.3), an LS estimate of $\mathbf{h}$ can be obtained by minimizing the error function $J = \mathbf{v}^T\mathbf{v}$. This yields

$$\hat{\mathbf{h}} = (\mathbf{U}^T\mathbf{U})^{-1}\mathbf{U}^T\mathbf{y} \qquad (7.4)$$

It has been shown in chapter 4 that $\hat{\mathbf{h}}$ is unbiased and consistent. However, $\hat{\mathbf{h}}$ does not have minimum variance because $\mathbf{v}$ is assumed autocorrelated. Therefore, the degree of correlation of $\mathbf{v}$ has a direct effect on the accuracy of $\hat{\mathbf{h}}$ for finite data length N.

Stage 2. In this stage, we make use of $u(k)$ and $\hat{\mathbf{h}}$ to generate an output

$\hat{w}(k)$ that is an estimate of the true system output $w(k)$, and then use the exact system equation

$$A(q^{-1})w(k) = B(q^{-1})u(k) \qquad (7.5)$$

to estimate the system parameters a_i and b_i. From equation (7.2), the estimated output $\hat{w}(k)$ can be computed by

$$\hat{w}(k) = \sum_{i=0}^{p} \hat{h}_i u(k-i) \qquad (7.6)$$

Then we can set up the approximate model

$$A(q^{-1})\hat{w}(k) = B(q^{-1})u(k) + \eta(k)$$

where $\eta(k)$ is the effective error resulted from the substitution of $\hat{w}(k)$ in equation (7.5). Now apply LS estimation to the above model to yield

$$\hat{\boldsymbol{\theta}} = (\hat{\mathbf{X}}^T \hat{\mathbf{X}})^{-1} \hat{\mathbf{X}}^T \hat{\mathbf{w}} \qquad (7.7)$$

where

$$\boldsymbol{\theta} = [a_1, \ldots, a_n, b_0, \ldots, b_n]^T$$
$$\hat{\mathbf{w}} = [\hat{w}(n+1), \hat{w}(n+2), \ldots, \hat{w}(n+N)]^T$$

$$\hat{\mathbf{X}} = \begin{bmatrix} -\hat{w}(n), \ldots, & -\hat{w}(1) & \vdots & u(n+1), \ldots, u(1) \\ \vdots & & & \\ -\hat{w}(n+N-1), \ldots, & -\hat{w}(N) & \vdots & u(n+N), \ldots, u(N) \end{bmatrix}$$

We see that as $N \to \infty$, $\hat{\mathbf{h}} \to \mathbf{h}$, $\hat{\mathbf{w}} \to \mathbf{w}$, $\eta \to 0$, $\hat{\boldsymbol{\theta}} \to \boldsymbol{\theta}$. Therefore $\hat{\boldsymbol{\theta}}$ is a consistent estimate of $\boldsymbol{\theta}$.

Stage 3. This stage is aimed at estimating the noise parameters in the following noise model (see figure 7-2):

$$C(q^{-1})\epsilon(k) = e(k) \qquad (7.8)$$

where

$$C(q^{-1}) = 1 + c_1 q^{-1} + \ldots + c_m q^{-m}$$

and the order m is assumed to be preselected. This can be easily accomplished by first computing the estimates $\hat{\epsilon}(k)$ from

$$\hat{\epsilon}(k) = y(k) + \sum_{i=1}^{n} \hat{a}_i y(k-i) - \sum_{i=0}^{n} \hat{b}_i u(k-i)$$

and then setting up the equation

$$C(q^{-1})\hat{\epsilon}(k) = \tilde{e}(k) \qquad (7.9)$$

where $\tilde{e}(k) = e(k) + \zeta(k)$, and $\zeta(k)$ is the effective error in the model because of the substitution of $\hat{\epsilon}(k)$. This yields the following LS estimates for the noise parameters:

$$\hat{c} = (\hat{\Omega}^T\hat{\Omega})^{-1}\hat{\Omega}^T\hat{\epsilon} \qquad (7.10)$$

where

$$c = [c_1, c_2, \ldots, c_m]^T$$

$$\hat{\epsilon} = [\hat{\epsilon}(m+1), \hat{\epsilon}(m+2), \ldots, \hat{\epsilon}(N+m)]^T$$

$$\hat{\Omega} = \begin{bmatrix} -\hat{\epsilon}(m), -\hat{\epsilon}(m-1), \ldots, & -\hat{\epsilon}(1) \\ & \vdots \\ -\hat{\epsilon}(N+m-1), \ldots, \ldots, & -\hat{\epsilon}(N) \end{bmatrix}$$

$\hat{c}$ is a consistent estimator of c because as $N \to \infty$, $\hat{\theta} \to \theta, \hat{\epsilon} \to \epsilon, \zeta \to 0$, and $\hat{c} \to c$.

Remark 1. The above three stages provide the complete solution to the system-noise identification problem. Recursive algorithms for each stage can be easily obtained if necessary (Pandya and Pagurek 1973). These algorithms are not presented here since they are identical to those in the previous chapters.

Another point we wish to make here is that the computation of $\hat{h}$ in stage 1 can be simplified if $\{u(k)\}$ is a pseudo-random binary sequence (PRBS). Refer to chapter 4 for detailed discussion.

Remark 2. The noise transfer function shown in figure 7-2 is $1/A(z^{-1})C(z^{-1})$. This transfer function can have a very high order $n + m$. In many practical system design situations, it is desirable to approximate this high-order transfer function by one of lower order in the form of $D(z^{-1})/\bar{C}(z^{-1})$, where

$$\bar{C}(z^{-1}) = 1 + \bar{c}_1 z^{-1} + \ldots + \bar{c}_f z^{-f}$$

$$D(z^{-1}) = 1 + d_1 z^{-1} + \ldots + d_f z^{-f}$$

We often require the order f to be $f < (1/2)(n + m)$. We discuss below how to estimate the parameters $\bar{c}_i$ and d_i when f is prespecified.

The new noise transfer function is shown in figure 7–3. This yields the relationship

$$\bar{C}(q^{-1})v(k) = D(q^{-1})e(k) \tag{7.11}$$

in which an estimate of $v(k)$ can be computed by

$$\hat{v}(k) = y(k) - \hat{w}(k)$$

with $\hat{w}(k)$ defined in equation (7.6). However, it is also necessary to get hold of $e(k)$ before the parameters can be estimated. It turns out that an estimate of $e(k)$ can be generated by

$$\hat{e}(k) = \hat{A}(q^{-1})\hat{C}(q^{-1})\hat{v}(k)$$

Then it follows from equation (7.11)

$$\hat{v}(k) = \sum_{i=1}^{f}[-\bar{c}_i\hat{v}(k-i) + d_i\hat{e}(k-i)] + \tilde{e}(k)$$

where $\tilde{e}(k) = \hat{e}(k) + \rho(k)$, and $\rho(k)$ is the effective error caused by substitution of $\hat{v}(k)$ and $\hat{e}(k)$. This equation allows us to obtain LS estimates for the noise parameters as follows:

$$\hat{\boldsymbol{\beta}} = (\boldsymbol{\Sigma}^T\boldsymbol{\Sigma})^{-1}\boldsymbol{\Sigma}^T\mathbf{v}$$

where

$$\boldsymbol{\beta} = [\bar{c}_i, \ldots, \bar{c}_f, d_1, \ldots, d_f]^T$$

$$\mathbf{v} = [\hat{v}(f+1), \ldots, \hat{v}(f+N)]^T$$

$$\boldsymbol{\Sigma} = \begin{bmatrix} -\hat{v}(f), & \ldots, & -\hat{v}(1), & \hat{e}(f), & \ldots, & \hat{e}(1) \\ \vdots & & & & & \\ -\hat{v}(N+f-1), & \ldots, & -\hat{v}(N), & \hat{e}(N+f-1), & \ldots, & \hat{e}(N) \end{bmatrix}$$

7.3 MSLS Method II

This method also uses the weighting sequence as an auxiliary model in stage 1. The noise model identification in stage 3 of method I is also applicable. The only difference between methods I and II is in stage 2. Here we take a different approach to estimate the system parameters from the weighting sequence (Iserman et al. 1974).

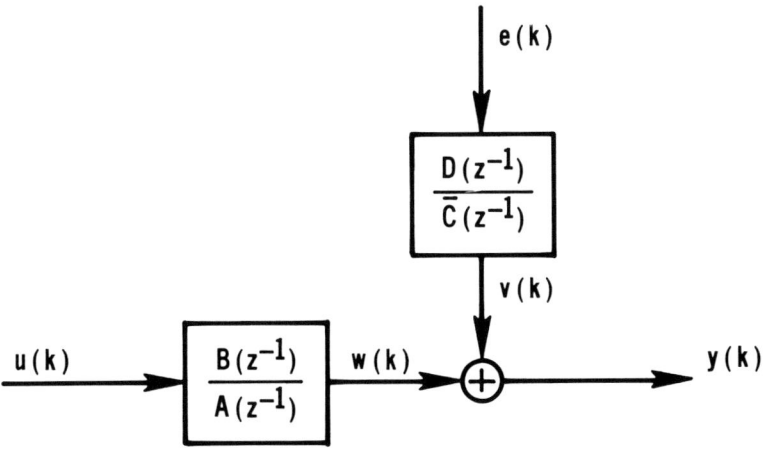

Figure 7-3. An Alternative Noise Transfer Function Model Frequently Used in System Identification

Stage 2. The definition of weighting sequence says that when a relaxed system is excited by a Kronecker delta function at $t = 0$, the response is the system's weighting sequence. Thus we have the following input-output data sequences (for $k \geq 0$):

$$\{u(k)\} = \{1, 0, 0, \ldots\}$$
$$\{w(k)\} = \{h_0, h_1, h_2, \ldots, h_p, 0, \ldots\}$$

Now apply these data to

$$A(q^{-1})w(k) = B(q^{-1})u(k)$$

and using $\hat{h}$ for h, we get

$$\begin{bmatrix} \hat{h}_0 \\ \hat{h}_1 \\ \vdots \\ \hat{h}_n \\ \hat{h}_{n+1} \\ \vdots \\ \hat{h}_p \end{bmatrix} = \begin{bmatrix} 0 & 0 & \cdots & 0 & | & 1 & 0 & \cdots & 0 \\ -\hat{h}_0 & 0 & \cdots & 0 & | & 0 & 1 & \cdots & 0 \\ \vdots & \vdots & & \vdots & | & \vdots & & & \vdots \\ -\hat{h}_{n-1} & -\hat{h}_{n-2} & \cdots & -\hat{h}_0 & | & 0 & \cdots & 1 & 0 \\ -\hat{h}_n & -\hat{h}_{n-1} & \cdots & -\hat{h}_1 & | & 0 & \cdots & 0 & 1 \\ \vdots & \vdots & & \vdots & | & \vdots & & & \\ -\hat{h}_{p-1} & -\hat{h}_{p-2} & \cdots & -\hat{h}_{p-n} & | & 0 & \cdots & & 0 \end{bmatrix} \begin{bmatrix} a_1 \\ a_2 \\ \vdots \\ a_n \\ b_0 \\ \vdots \\ b_n \end{bmatrix} + \begin{bmatrix} \eta_0 \\ \eta_1 \\ \vdots \\ \eta_n \\ \vdots \\ \eta_p \end{bmatrix}$$

where η is the random error associated with the substitution of $\hat{h}$. Denote the above by

$$\hat{\mathbf{h}} = \hat{\mathbf{H}}\boldsymbol{\theta} + \boldsymbol{\eta} \qquad (7.12)$$

This immediately results in a LS estimate $\hat{\boldsymbol{\theta}}$

$$\hat{\boldsymbol{\theta}} = (\hat{\mathbf{H}}^T\hat{\mathbf{H}})^{-1}\hat{\mathbf{H}}^T\hat{\mathbf{h}} \qquad (7.13)$$

Remark. We note that one basic condition for $\hat{\boldsymbol{\theta}}$ to exist is that $p \geq 2n + 1$. However, in practice, it is undesirable to have p too large so as to cause the computation of $\hat{\mathbf{h}}$ to be too costly. Thus the smoothing power of the above LS procedure is limited because of the limited number of p equations in equation (7.12). This implies that the estimate $\hat{\boldsymbol{\theta}}$ is not as accurate as that of method I. Still, $\hat{\boldsymbol{\theta}}$ is a consistent estimator because as the data length $N \to \infty$, $\hat{\mathbf{h}} \to \mathbf{h}$, and $\hat{\boldsymbol{\theta}} \to \boldsymbol{\theta}$.

7.4 MSLS Method III

This method is completely different from the previous two methods in all three stages. Instead of using the weighting sequence as auxiliary model, we use an "enlarged" difference equation in which the residuals become uncorrelated when fitted to the system's input-output data. This enlarged system is then identified by least squares. Subsequently, we estimate the original system parameters and the noise parameters in two separate LS stages.

To clarify what we mean by enlarged system, let us combine equations (7.1) and (7.8) to write

$$C(q^{-1})A(q^{-1})y(k) = C(q^{-1})B(q^{-1})u(k) + e(k) \qquad (7.14)$$

suppose we define two new polynomials $E(q^{-1})$ and $F(q^{-1})$

$$E(q^{-1}) = C(q^{-1})A(q^{-1}), \quad F(q^{-1}) = C(q^{-1})B(q^{-1}) \qquad (7.15)$$

then

$$E(q^{-1})y(k) = F(q^{-1})u(k) + e(k) \qquad (7.16)$$

This enlarged system (auxiliary model) is of order $(m + n)$, and the residual $e(k)$ is white. The polynomials $A(q^{-1})$, $B(q^{-1})$, and $C(q^{-1})$ are related to the auxiliary model through equation (7.15). We now present below the details of the MSLS solution procedures (Hsia 1975).

Stage 1. This stage is aimed at estimating the auxiliary model of equation (7.16). Let

$$E(q^{-1}) = 1 + e_1 q^{-1} + e_2 q^{-2} + \ldots + e_{m+n} q^{-(m+n)}$$
$$F(q^{-1}) = f_0 + f_1 q^{-1} + f_2 a^{-2} + \ldots + f_{m+n} q^{-(m+n)}$$
(7.17)

Define the parameter vector α as

$$\alpha = [e_1, \ldots, e_{m+n}, f_0, \ldots, f_{m+n}]^T$$

The LS estimate of α from equation (7.16) is

$$\hat{\alpha} = (\Phi^T \Phi)^{-1} \Phi^T y \qquad (7.18)$$

where

$$y = [y(m + n + 1), \ldots, y(m + n + N)]^T$$

$$\Phi = \begin{bmatrix} -y(m+n), \ldots, & -y(1), & u(m+n+1), \ldots, u(1) \\ \vdots & & \vdots \\ -y(m+n+N-1), \ldots, & -y(N), & u(m+n+N), \ldots, u(N) \end{bmatrix}$$

Stage 2. From equation (7.15), we see that

$$B(q^{-1}) E(q^{-1}) = A(q^{-1}) F(q^{-1}) \qquad (7.19)$$

with $E(q^{-1})$ and $F(q^{-1})$ being estimated through $\hat{\alpha}$. We can use equation (7.19) to estimate $A(q^{-1})$ and $B(q^{-1})$. To implement this procedure, we multiply out equation (7.19), and equate the coefficients of the like powers of q on both sides. This leads to a set of $2n + m + 1$ linear equations of a_i and b_i. Now replace all e_i and f_i terms by their estimates, and express these equations in the following vector form:

$$g(\hat{e}, \hat{f}) = G(\hat{e}, \hat{f})\theta + \eta \qquad (7.20)$$

where g and G are respectively $(2n + m + 1) \times 1$ and $(2n + m + 1) \times (2n + 1)$ matrices with $\hat{e}_i$ and $\hat{f}_i$ as elements, η is the random error vector in the equation. Thus, the LS estimate of θ can be immediately expressed as

$$\hat{\theta} = (G^T G)^{-1} G^T g \qquad (7.21)$$

Stage 3. To estimate the noise polynomial $C(q^{-1})$, we can follow the procedures detailed in stage 3 of method I. However, we propose here an alternative approach that is based on the identities in equation (7.15). These identities provide $2(m + n) + 1$ linear equations for the parameters

c_i when coefficients of like powers of q on both sides of the identities are equated. In analogy to stage 2, we can set up the vector equation

$$\mathbf{r}(\hat{\alpha}, \hat{\theta}) = \mathbf{R}(\hat{\alpha}, \hat{\theta})\mathbf{c} + \boldsymbol{\zeta} \tag{7.22}$$

where $\hat{\alpha}$ and $\hat{\theta}$ result from the substitution of $\hat{A}(q^{-1})$, $\hat{B}(q^{-1})$, $\hat{E}(q^{-1})$, $\hat{F}(q^{-1})$ in equation (7.15), and $\boldsymbol{\zeta}$ is the associated equation error vector. Thus we obtain the LS estimate $\hat{\mathbf{c}}$ as

$$\hat{\mathbf{c}} = (\mathbf{R}^T\mathbf{R})^{-1}\mathbf{R}^T\mathbf{r} \tag{7.23}$$

Remark 1. It is easy to see that $\hat{\theta}$ is consistent because $\hat{\alpha}$ is consistent. Consistency of both $\hat{\alpha}$ and $\hat{\theta}$ also insure the consistency of $\hat{\mathbf{c}}$.

This method is a more direct approach for parameter estimation than the other two methods. In addition, the use of weighting sequence tends to create a very high dimension LS problem that is more computationally demanding. Equation (7.16) shows that a white residual can result if we fit the input-output data by a model with higher enough order. However, this model has common factors in its transfer function. The true system transfer function can be obtained if the common factors are removed (Van den Boom and Van Den Eden 1974). This is essentially the task of stage 2.

Remark 2. To demonstrate how to set up the LS equations in steps 2 and 3, we consider as an example the following simple system:

$$A(q^{-1}) = 1 + a_1 q^{-1}, \quad B(q^{-1}) = b_0, \quad C(q^{-1}) = 1 + c_1 q^{-1}.$$

This yields

$$E(q^{-1}) = C(q^{-1})A(q^{-1}) = 1 + (a_1 + c_1)q^{-1} + a_1 c_1 q^{-2}$$
$$= 1 + e_1 q^{-1} + e_2 q^{-2}$$
$$F(q^{-1}) = C(q^{-1})B(q^{-1}) = b_0 + b_0 c_1 q^{-1} = f_0 + f_1 q^{-1}$$
$$A(q^{-1})F(q^{-1}) = f_0 + (a_1 f_0 + f_1)q^{-1} + a_1 f_1 q^{-2}$$
$$B(q^{-1})E(q^{-1}) = b_0 + b_0 e_1 q^{-1} + b_0 e_2 q^{-2}$$

From the last two expressions, we can derive dequation (7.20) in stage 2 as

$$\begin{bmatrix} \hat{f}_0 \\ \hat{f}_1 \\ 0 \end{bmatrix} = \begin{bmatrix} 0 & 1 \\ -\hat{f}_0 & \hat{e}_1 \\ -\hat{f}_1 & \hat{e}_2 \end{bmatrix} \begin{bmatrix} a_1 \\ b_0 \end{bmatrix} + \begin{bmatrix} \eta_1 \\ \eta_2 \\ \eta_3 \end{bmatrix}$$

Also from the expressions of $E(q^{-1})$ and $F(q^{-1})$, we find equation (7.22) in stage 3 to be

$$\begin{bmatrix} \hat{e}_1 - \hat{a}_1 \\ \hat{e}_2 \\ \hat{f}_1 \end{bmatrix} = \begin{bmatrix} 1 \\ \hat{a}_1 \\ \hat{b}_0 \end{bmatrix} [c_1] + \begin{bmatrix} \zeta_1 \\ \zeta_2 \\ \zeta_3 \end{bmatrix}$$

Remark 3. The three LS solutions in this method can be converted to recursive algorithms (Hsia 1975). The conversion in stage 1 is straightforward, and we will not discuss it here. However, the conversion of stages 2 and 3 is different in that the recursion is not for the increase of observed data, but rather for updating $\hat{\theta}$ and $\hat{c}$ subsequent to the updating of $\hat{\alpha}$ in stage 1. We now derive these two recursive algorithms.

Consider that $\hat{\alpha}(N)$ is updated to $\hat{\alpha}(N + 1)$ in stage 1, and $\hat{\alpha}(N + 1) = \hat{\alpha}(N) + \Delta_\alpha$. Then, in stage 2 we can write

$$G(N + 1) = G(N) + \Delta_G \quad g(N + 1) = g(N) + \Delta_g$$

Define $\Lambda = (G^T G)^{-1}$; then

$$\Lambda(N + 1) = [\Lambda^{-1}(N) + \Delta_\Lambda]^{-1} \cong \Lambda(N) - \Lambda(N)\Delta_\Lambda \Lambda(N)$$

where

$$\Delta_\Lambda = \Delta_G^T [G(N) + \Delta_G] + [G(N) + \Delta_G]^T \Delta_G + \Delta_G^T \Delta_G$$

Substituting $\Lambda(N + 1)$ and $g(N + 1)$ into equation (7.21), we get

$$\hat{\theta}(N + 1) = \hat{\theta}(N) + \Lambda(N)[\Delta_g - \Delta_\Lambda \hat{\theta}(N) - \Delta_\Lambda \Lambda(N)\Delta_g]$$

Upon neglecting the higher-order terms of Δ and simplifying, we get the desired recursive algorithm

$$\hat{\theta}(N + 1) = \hat{\theta}(N) + \Lambda(N)[\Delta_g - (\Delta_G^T G(N) + G^T(N)\Delta_G)\hat{\theta}(N)]$$

$$\Lambda(N + 1) = \Lambda(N) - \Lambda(N)[\Delta_G^T G(N) + G^T(N)\Delta_G]\Lambda(N)$$

We note that Δ_G and Δ_g are functions of $\hat{\alpha}(N)$ and Δ_α.

In a similar manner, we can derive the following recursive algorithm for equation (7.23):

$$\hat{c}(N + 1) = \hat{c}(N) + \Psi(N)[\Delta_r - (\Delta_R^T R(N) + R^T(N)\Delta_R)\hat{c}(N)]$$

$$\Psi(N + 1) = \Psi(N) - \Psi(N)[\Delta_R^T R(N) + R^T(N)\Delta_R]\Psi(N)$$

where

$$\Delta_R = R(N + 1) - R(N) \quad \Delta_r = r(N + 1) - r(N)$$

7.5 Comparison of the MSLS and GLS Methods

In this section, we present computer-simulated results comparing the estimation accuracies and computer time requirements for the three MSLS methods just described. Also included in the comparison is the GLS algorithm. This comparison study, although limited in scope, is intended to show that MSLS generally uses less computer time than does GLS, and that MSLS estimates are also less accurate.

The following third-order system is chosen for study:

$$A(q^{-1}) = 1 + 0.90q^{-1} + 0.15q^{-2} + 0.02q^{-3}$$
$$B(q^{-1}) = 0.70q^{-1} - 1.5q^{-2} \quad (7.24)$$
$$C(q^{-1}) = 1 + 1.0q^{-1} + 0.41q^{-2}$$

The computer simulation is implemented according to the following conditions.

1. Both the input $u(k)$ and the noise $e(k)$ are independent Gaussian random variables of zero mean value. The signal-to-noise ratio at the output is set at $\sigma_w^2/\sigma_v^2 = 1.18$.

2. We truncate the weighting sequence $\{h_k\}$ after $k = 10$. The actual weighting sequence is shown in figure 7–4.

3. The exact structures and orders of the system are used in the model to be identified. That is, the model has the following coefficient polynomials:

$$A(q^{-1}) = 1 + a_1 q^{-1} + a_2 q^{-2} + a_3 q^{-3}$$
$$B(q^{-1}) = b_1 q^{-1} + b_2 q^{-2} \quad (7.25)$$
$$C(q^{-1}) = 1 + c_1 q^{-1} + c_2 q^{-2}$$

Identifications are performed for two different data lengths, $N = 300$ and $N = 450$. The resulting parameter estimates are shown in tables 7–1 and 7–2. We now make the following observations.

1. The GLS algorithm converged after six iterations for both cases. We can see that the MSLS methods used 30 to 60 percent less computer time than did the GLS algorithm. In general, the computer time for all methods increases as N increases.

2. Among the MSLS methods, method III requires the least amount of computer time. Methods I and II rank second and third although the difference between them is small. The major cause for the time differential lies in stage 1. In methods I and II, the LS solution requires the inversion of an 11×11 matrix, whereas in method III, the matrix being inverted is 9×9. This matrix dimensional difference contributes to the lesser compu-

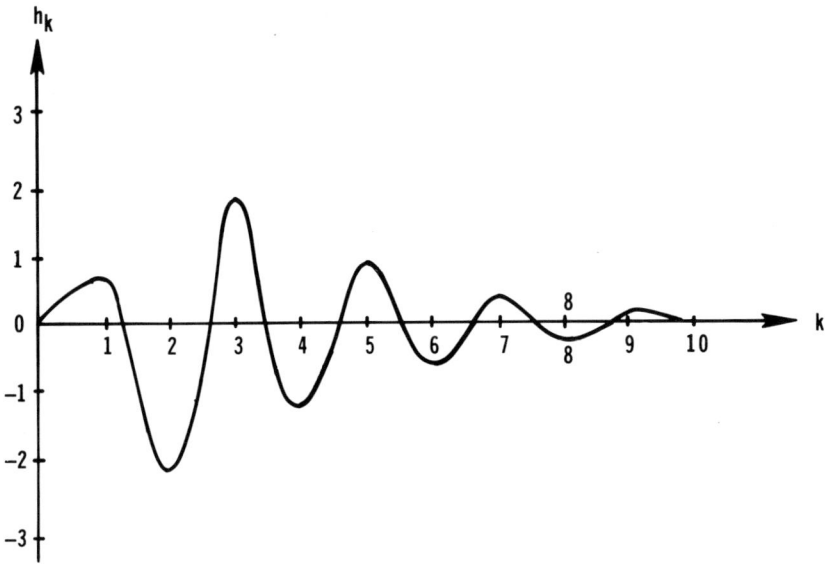

Figure 7-4. Weighting Sequence $\{h_k\}$ of a Third-order Linear System

Table 7-1
Identification Results of a Third-order System, Data Length $N = 300$, Signal/Noise $= 1.18$, MSE $=$ Mean-square-error, Mean and MSE Computed from 10 Estimation Runs

True values	Parameter estimates	MSLS methods			GLS method
		I	II	III	
a_1	Mean	0.90470	0.91232	0.89584	0.89277
(0.90)	MSE	0.00121	0.00131	0.00118	0.00163
a_2	Mean	0.16998	0.17103	0.17619	0.17993
(0.15)	MSE	0.00498	0.00520	0.00255	0.00250
a_3	Mean	0.00080	0.00198	0.02480	0.01958
(0.02)	MSE	0.00414	0.00469	0.00167	0.00155
b_1	Mean	0.72162	0.71879	0.70056	0.70047
(0.70)	MSE	0.00545	0.00332	0.00233	0.00262
b_2	Mean	-1.48753	-1.48488	-1.48800	-1.48706
(-1.5)	MSE	0.00396	0.00339	0.00307	0.00322
c_1	Mean	Not	Not	0.99389	1.00690
(1.0)	MSE	computed	computed	0.00662	0.00755
c_2	Mean	Not	Not	0.38777	0.37925
(0.41)	MSE	computed	computed	0.00472	0.00662
Computing time		1.84 min	1.48 min	1.12 min	2.76 min

Table 7-2
Identification Results of a Third-order System, Data Length $N = 450$, Signal/Noise = 1.18, MSE = Mean-square-error, Mean and MSE Computed from 10 Estimation Runs

True values	Parameter estimates	MSLS methods			GLS method
		I	II	III	
a_1 (0.90)	Mean MSE	0.89156 0.00067	0.90041 0.00062	0.89558 0.00070	0.90108 0.00054
a_2 (0.15)	Mean MSE	0.16071 0.00061	0.16390 0.00073	0.15798 0.00071	0.15448 0.00066
a_3 (0.02)	Mean MSE	0.032605 0.00059	0.04276 0.00143	0.01463 0.00057	0.02478 0.00018
b_1 (0.70)	Mean MSE	0.69620 0.00250	0.68157 0.00311	0.69688 0.00092	0.69910 0.00084
b_2 (−.15)	Mean MSE	−1.50061 0.00149	−1.49250 0.00199	−1.49700 0.00112	−1.50039 0.00089
c_1 (1.0)	Mean MSE	Not computed	Not computed	1.01149 0.00476	1.00875 0.00447
c_2 (0.41)	Mean MSE	Not computed	Not computed	0.40693 0.00214	0.41115 0.00187
Computing time		2.98 min	2.45 min	1.72 min	4.62 min

tation time of method III. Furthermore, in stage 2 of method I, the number of multiplications is larger than that of method II ($N \gg \rho$). Therefore, method I is somewhat more costly to compute.

3. In terms of parameter estimates accuracy, the GLS algorithm in both cases gives the smallest mean square error (MSE). Thus, as expected, GLS is the most accurate method. The results also show that the MSE decreases as N increases for all four methods.

Among the MSLS methods, the accuracy of method III is better than that of the other two methods. In fact, method III comes very close to the GLS method. The accuracies of methods I and II are comparable.

4. When taking into account both the computational simplicity and the estimation accuracy, we regard MSLS method III to be an excellent system identification technique.

7.6 Concluding Remarks

In this chapter, we have presented three LS methods called MSLS. It is shown that the MSLS methods can yield consistent system parameter estimates in two separate stages, each of which involves only a simple LS.

An additional LS stage is executed if noise parameters are also required. This is quite different from the GLS method of chapter 6, which requires iterative solution procedures. However, the computational simplicity achieved by the MSLS approach is at the expense of lesser estimation accuracy. These facts have been clearly demonstrated by a computer-simulated example in which all three MSLS estimates were computed along with the GLS estimates. The comparison has led to the conclusion that the MSLS method III is a method with good overall performance when both computer-time cost and estimation accuracy are taken into consideration. Additional comparison studies involving other identification techniques can be found in Iserman, et al. (1974), Sardis (1974) and Gustovsson (1972).

References

Gustovsson, I., "Comparison of Different Methods for Identification of Industrial Processes," *Automation*, Vol. 8, pp. 127–142, Mar. 1972.

Hsia, T. C., "On Multistage Least Squares Approach to Systems Identification," *Proceedings, IFAC Sixth World Congress*, Paper 18.2, Boston, 1975.

Isermann, R.; Baur, U,; Bamberger, W.; Kneppo, P.; and Siebert, H.; "Comparison of Six On-line Identification and Parameter Estimation Methods," *Automatica*, Vol. 10, pp. 81–103, Jan. 1974.

Pandya, R. N., and Pagurek, B., "Two Stage Least Squares Estimators and Their Recursive Approximations," 3d IFAC *Symposium on Identification and System Parameter Estimation*, pp. 701–710, Hague, June 1973.

Saridis, G. N., "Comparison of Six On-line Identification Algorithms," *Automatica*, Vol. 10, pp. 69–79, Jan. 1974.

Van den Boom, A. J., W., and Van Den Enden, A. W. M., "The Determination of the Orders of Process and Noise Dynamics," *Automation*, Vol. 10, pp. 245–256, May 1974.

8 Identification of Nonlinear Systems

8.1 Introduction

In this last chapter we wish to discuss the identification problem of nonlinear discrete-time systems. Because of space limitations, we will only cover those types of system structures that can be easily identified by least-squares methods.

One of the major difficulties in dealing with nonlinear systems is the lack of unified mathematical theory for representing various nonlinear system characteristics. Thus it is not practical to talk about identification of nonlinear systems unless a specific system representation is imposed beforehand. In this chapter, we are mainly interested in identifying unknown parameters of preselected structures in nonlinear systems.

There are a number of representations for nonlinear systems that are suitable for system identification purposes. The simplest is that in which system is characterized by nonlinear differential (or difference) equations. Another well-known representation is the Volterra series, which is good for describing the input-output behavior of a nonlinear system. A wide class of nonlinear systems can also be represented by the Hammerstein model, which is composed of a nonlinear gain followed by a linear subsystem. We will discuss in detail the identification of each of these system representations.

8.2 Volterra Series Representation and Identification

Consider the single input-single output nonlinear system shown in figure 8–1. The input-output relationship of this system can be expressed explicitly as the Volterra series (Volterra 1959; Eykhoff 1974; Schetzen 1974):

$$w(t) = \int_{-\infty}^{t} g_1(\tau)u(t-\tau)\,d\tau + \int_{-\infty}^{t} g_2(\tau_1, \tau_2)u(t-\tau_1)u(t-\tau_2)\,d\tau_1 d\tau_2$$

$$+ \ldots + \int_{-\infty}^{t} \ldots \int_{-\infty}^{t} g_n(\tau_1 \ldots \tau_n) \prod_{i=1}^{n} u(t-\tau_i)\,d\tau_i + \ldots \quad (8.1)$$

The nth order Volterra Kernel $g_n(\tau_1 \ldots \tau_n)$ represents the weighting function of nth degree. Thus the nth-order term is an n-fold convolution inte-

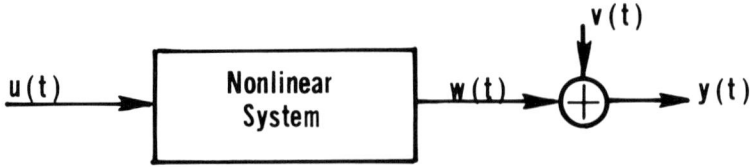

Figure 8–1. Nonlinear Dynamic System with Output Additive Noise

gral. It is obvious that linear systems are treated by Volterra series representation as a special case.

Assume that the nonlinear system in question is stable with finite settling time. To identify the Kernels $g_n(\tau_1 \ldots \tau_n)$, we can simply follow the procedures described in chapter 4. That is, we approximate the Volterra series by its sampled-data form

$$w(k) = \sum_{i=0}^{p} h(i)u(k-i) + \sum_{i=0}^{p}\sum_{j=0}^{p} h(i,j)u(k-i)u(k-j)$$

$$+ \sum_{i=0}^{p}\sum_{j=0}^{p}\sum_{m=0}^{p} h(i,j,m)u(k-i)u(k-j)u(k-m) + \ldots \quad (8.2)$$

for $k \geq p$. In the above we have used T as the sampling time where $t_k = kT$, pT is the settling time, $h(i) = g_1(\tau = iT)T$, $h(i,j) = g_2(\tau_1 = iT, \tau_2 = jT)T^2$, and so on. We note that T is dropped in all the arguments for the sake of simplicity.

The objective now is to estimate $h(i)$, $h(i,j)$, and so on from $u(k)$ and $y(k)$ data sequences (Roy and Sherman 1967). This estimation can be easily accomplished by the least-squares technique described in chapter 4. For the purpose of illustration, let the Volterra series in equation (8.2) be truncated after the quadratic terms. Then we can write

$$y(k) = \boldsymbol{\mu}^T(k)\mathbf{h} + v(k) \quad (8.3)$$

where $v(k)$ is the random additive noise

$$\boldsymbol{\mu}(k) = [u(k), \ldots, u(k-p) \vdots u^2(k), u(k)u(k-1), \ldots, u^2(k-p)]^T$$

$$\mathbf{h} = [h(0), \ldots, h(p) \vdots h(0,0), h(0,1), \ldots, h(p,p)]^T$$

We see that by mimizing the cost function $J = \Sigma_k v^2(k)$, the LS estimate of $\mathbf{h}$ can be obtained from equation (8.3) for $k = 1, 2, 3, \ldots, N$ ($N >$ dimension of $\mathbf{h}$). Details of the solution of $\hat{\mathbf{h}}$ and its properties can be found in chapter 4.

8.3 Nonlinear Difference Equations with Linear Parameters

Like the weighting sequence of linear systems, the Volterra series is a nonparametric representation of nonlinear systems. The identification of this input-output model always leads to a high-dimensional parameter vector **h**. Parametric model representations such as nonlinear difference (or differential) equations usually require a much smaller number of parameters to be identified. However, the use of nonlinear difference equations requires detailed knowledge of the internal system structures.

There are two ways that parameters can appear in nonlinear difference equations: linearly or nonlinearly. Identification of the linear case is much more straightforward than the nonlinear case. The Volterra series in equation (8.3) is a special type of linear-parameter identification problem in that nonlinearity only involves the input $u(k)$.

In general, a nonlinear difference equation can be written as

$$w = f(w, u, \boldsymbol{\theta}) \tag{8.4}$$

Given below is an example of a linear-parameter nonlinear difference equation:

$$w(k) = \theta_1 w^2(k-1) + \theta_2 u(k-1) + \theta_3 u(k-1)w(k)$$

We see that there are both linear and nonlinear terms in $u(k)$ and $w(k)$, but the parameters θ_i are strictly linear. Identifying $\boldsymbol{\theta}$ in this case then is the same as in linear difference equations.

Replacing $w(k)$ by the noise-corrupted output $y(k)$, and defining

$$\boldsymbol{\theta} = [\theta_1, \theta_2, \theta_3]^T$$

$$\mathbf{x}(k) = [y^2(k-1), u(k-1), u(k-1)y(k)]^T$$

We can write

$$y(k) = \mathbf{x}^T(k)\boldsymbol{\theta} + \epsilon(k) \tag{8.5}$$

where $\epsilon(k)$ is the equation error. Now $\boldsymbol{\theta}$ can be estimated by least squares from equation (8.5).

We observe that $\epsilon(k)$ is a nonlinear function of $u(k)$, $y(k)$, and $v(k)$:

$$\epsilon(k) = \theta_1[v^2(k-1) - 2y(k-1)v(k-1)] + \theta_3 u(k-1)v(k) + v(k)$$

Thus the LS estimate $\hat{\boldsymbol{\theta}}$ is biased when the noise $v(k)$ is present. The complicated nature of $\epsilon(k)$ even makes the GLS method inapplicable. Therefore, it is difficult in general to obtain good LS parameter estimates in nonlinear systems when there are significant noises present. Under the assumption that only the output data is noise-corrupted, it is possible to obtain consistent LS parameter estimates when the nonlinear difference equation contains only input nonlinear terms. A class of this nonlinear systems is called the Hammerstein model, which is discussed in section 8.5

Application of the identification procedures outlined in this section to experimental biological data modeling can be found in Inbar, Hsia, and Baskin (1970).

8.4 Nonlinear Difference Equations with Nonlinear Parameters

A more difficult identification problem occurs when the system parameters appear in nonlinear forms in equation (8.4). An example of such a case is found in Hsia (1976a),

$$w(k) = \theta_1 \sin(\theta_2 k + \theta_3) \tag{8.6}$$

Here θ_1 is the amplitude of a sinusoidal function, θ_2 is the angular frequency, and θ_3 is the phase angle. Estimating θ_i from data $\{w(k), k\}$ cannot be accomplished by the usual LS procedure. Rather an iterative algorithm is required. We will now describe below such an identification algorithm.

Referring to equation (8.4), we wish to estimate $\boldsymbol{\theta}$ by minimizing the error function

$$J = \sum_{k=1}^{N} [w(k) - f(w(k), u(k), \boldsymbol{\theta})]^2$$

It is assumed that $(\partial J/\partial \boldsymbol{\theta}) = 0$ does not lead to linear equations of $\boldsymbol{\theta}$. Let $\hat{\boldsymbol{\theta}}(i)$ be the ith iteration estimate of $\boldsymbol{\theta}$, and $\hat{\boldsymbol{\theta}}(i + 1)$ be

$$\hat{\boldsymbol{\theta}}(i + 1) = \hat{\boldsymbol{\theta}}(i) + \Delta\boldsymbol{\theta}$$

The corrective term $\Delta\boldsymbol{\theta}$ is determined in such as way that $J[\Delta\boldsymbol{\theta}]$ is a minimum. This step can be easily accomplished by expanding $f(w, u, \boldsymbol{\theta}(i) + \Delta\boldsymbol{\theta})$ in the Taylor series about $\boldsymbol{\theta}(i)$

$$f(w, u, \boldsymbol{\theta}(i) + \Delta\boldsymbol{\theta}) \approx f(w, u, \boldsymbol{\theta}(i)) + f'(\boldsymbol{\theta}(i))\Delta\boldsymbol{\theta}$$

where

$$f'(\boldsymbol{\theta}(i)) = \left. \frac{\partial f}{\partial \boldsymbol{\theta}} \right|_{\boldsymbol{\theta}=\boldsymbol{\theta}(i)}$$

Then $J[\Delta\theta]$ becomes

$$J[\Delta\theta] = \sum_{k=1}^{N} [w(k) - f(\boldsymbol{\theta}(i)) - f'(\boldsymbol{\theta}(i))\Delta\boldsymbol{\theta}]^2$$

Now $\Delta\boldsymbol{\theta}$ is determined by setting $(\partial J/\partial\Delta\boldsymbol{\theta}) = \mathbf{0}$.

Once $\Delta\boldsymbol{\theta}$ is calculated, $\hat{\boldsymbol{\theta}}(i+1)$ is then determined. This iterative process will lead to a $\hat{\boldsymbol{\theta}}$ that minimizes J. But J is a nonquadratic function of $\boldsymbol{\theta}$, which means that local minima may exist. Therefore, it is essential to select an initial value $\hat{\boldsymbol{\theta}}(0)$ that is close to the true $\boldsymbol{\theta}$ for the algorithm to converge to the correct parameter values.

As an example, let us consider equation (8.6) for the case that θ_1 and θ_3 are known quantities, and only θ_2 is to be estimated. We have (see Hisa 1976a)

$$J[\Delta\theta_2] = \sum_{k=1}^{N} [w(k) - \theta_1 \sin(\theta_2(i)k + \theta_3)$$
$$- k\theta_1 \cos(\theta_2(i) k + \theta_3)\Delta\theta_2]^2$$

This yields the following solution of $\Delta\theta_2$

$$\Delta\theta_2 = (\mathbf{H}^T\mathbf{H})^{-1}\mathbf{H}^T\mathbf{g}$$

where

$$\mathbf{H} = \begin{bmatrix} \theta_1 \cos(\theta_2(i) + \theta_3) \\ 2\theta_1 \cos(2\theta_2(i) + \theta_3) \\ \vdots \\ N\theta_1 \cos(N\theta_2(i) + \theta_3) \end{bmatrix}$$

$$\mathbf{g} = \begin{bmatrix} w(1) - \theta_1\sin(\theta_2(i) + \theta_3) \\ w(2) - \theta_1 \sin(2\theta_2(i) + \theta_3) \\ \vdots \\ w(N) - \theta_1 \sin(N\theta_2(i) + \theta_3) \end{bmatrix}$$

8.5 Hammerstein model—GLS Identification

There is a wide class of nonlinear systems that can be modelled by interconnected memoryless nonlinear gains and linear subsystems. Three basic configurations of such models are possible (Bányász, Haber, and Keviczky 1973): the Wiener model, the Hammerstein model, and the general model. These are depicted in figure 8–2.

Our concern in this section and the next is with the identification of the Hammerstein model. A detailed structure of this model is shown in figure 8–3. The nonlinear gain is approximated by a power polynomial of order p

$$x(k) = \gamma_1 u(k) + \gamma_2 u^2(k) + \ldots + \gamma_p u^p(k). \tag{8.7}$$

The coefficients γ_i and order p can be appropriately selected to approximate a given memoryless nonlinear gain. The linear subsystem has a difference equation of nth order

$$A(q^{-1})y(k) = B(q^{-1})x(k) \tag{8.8}$$

when

$$A(q^{-1}) = 1 + a_1 q^{-1} + \ldots + a_n q^{-n}$$

$$B(q^{-1}) = b_0 + b_1 q^{-1} + \ldots + b_n q^{-n}$$

The linear system is assumed to be stable. The output additive noise $v(k)$ is a random variable of zero mean. The identification problem is to estimate the parameters a_i, b_i, and γ_i, for preselected n and p, from the measured data sequence $\{u(k), y(k)\}$.

Before we set out to tackle the identification problem, let us examine some basic properties of the Hammerstein model. Combining equations (8.7), (8.8), we get the following overall system equation:

$$A(q^{-1})y(k) = B(q^{-1})\left[u(k) + \sum_{i=1}^{p} \gamma_i u^i(k) \right] + \epsilon(k) \tag{8.9}$$

where

$$\epsilon(k) = A(q^{-1})v(k) \tag{8.10}$$

Notice that without loss of generality, the nonlinear gain coefficient γ_1 is normalized to unity. We also notice that the remaining γ_i's will multiply the b_i's in $B(q^{-1})$ so that cross-products $\gamma_i b_j$ will result. This means that we have a nonlinear-parameter identification problem to solve. However, when these cross-products are taken as new parameters, the problem becomes linear. So the nonlinear-parameter estimation procedures outlined in section 8.4 need not be used. Nevertheless, additional steps must be taken to separate the parameters γ_i and b_j from $\gamma_i b_j$.

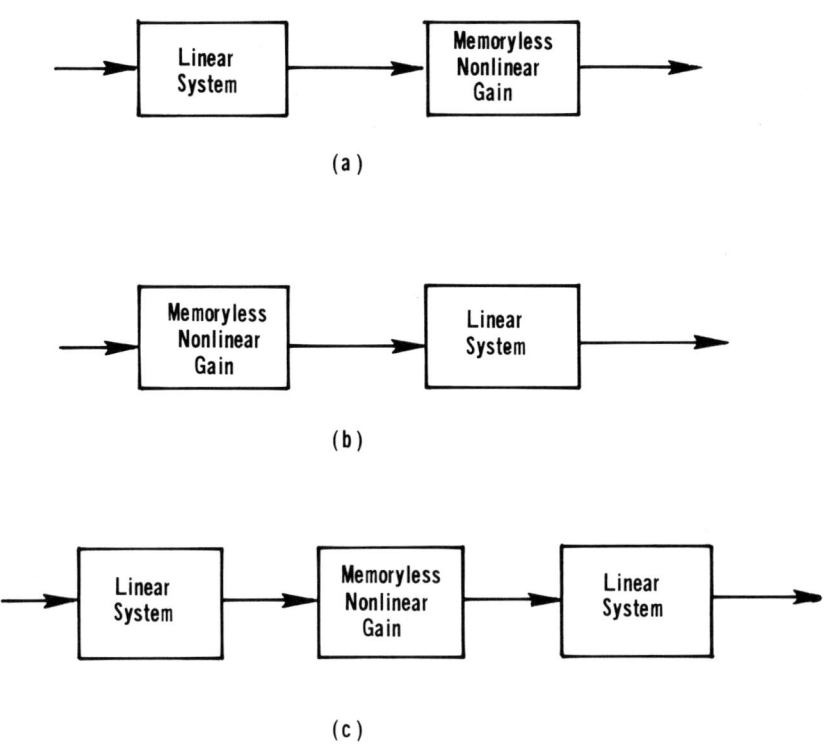

Figure 8–2. Three Types of Nonlinear System Models: (a) Wiener Model (b) Hammerstein Model (c) General Model

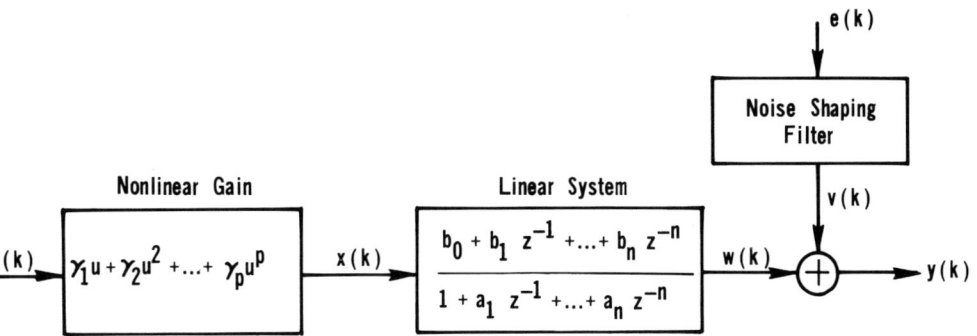

Figure 8–3. Hammerstein Model Structure

When the disturbing noise in the system is low, the simple least-squares method is adequate to identify the parameters (Hsia 1968; Chang and Luus 1971). As the noise becomes more significant, the LS method will yield biased parameter estimates. In this case we have to apply iterative LS methods (Narenda and Gallman 1966; Haist, Chang, and Luus 1973). In the following we present the generalized least-squares solution to the problem. First we define the polynomial $S_i(q^{-1})$:

$$S_i(q^{-1}) = \gamma_i B(q^{-1}) = s_{i0} + s_{i1}q^{-1} + \ldots + s_{in}q^{-n} \tag{8.11}$$

where

$$s_{ij} = \gamma_i b_j \quad i = 1, \ldots, p \quad j = 1, \ldots, n \tag{8.12}$$

We have

$$A(q^{-1})y(k) = B(q^{-1})u(k) + \sum_{i=2}^{p} S_i(q^{-1})u^i(k) + \epsilon(k) \tag{8.13}$$

We also assume that the residual $\epsilon(k)$ satisfy the autoregressive model

$$C(q^{-1})\epsilon(k) = e(k) \tag{8.14}$$

where $e(k)$ is an independent random variable of zero mean, and $C(q^{-1})$ is

$$C(q^{-1}) = 1 + c_1 q^{-1} + \ldots + c_m q^{-m} \tag{8.15}$$

in which c_i are unknown, and the order m is usually preselected. The GLS method is applied to estimate a_i, b_i, s_i, and c_i. γ_i are subsequently separated from s_i in one separate step.

Define a set of filtered signals

$$\tilde{u}^i(k) = C(q^{-1})u^i(k) \quad \tilde{y}(k) = C(q^{-1})y(k) \tag{8.16}$$

We can show that equation (8.13) becomes

$$A(q^{-1})\tilde{y}(k) = B(q^{-1})\tilde{u}(k) + \sum_{i=2}^{p} S_i(q^{-1})\tilde{u}^i(k) + e(k). \tag{8.17}$$

This equation and equation (8.14) are then alternately applied in the GLS algorithm. We now briefly describe the GLS algorithm. Detailed descriptions can be found in chapter 6.

Step 1. Let $c_i = 0$. Obtain the LS estimates of the parameter set $(\hat{a}_i, \hat{b}_i, \hat{s}_{ij})$ from equation (8.17) using the unfiltered data $u^i(k)$ and $y(k)$.

Step 2. Generate the residual sequence $\{\epsilon(k)\}$ from equation (8.13). Then obtain the LS estimates of c_i from equations (8.14), (8.15).

Step 3. Generate the filtered signals $\bar{y}(k)$ and $\bar{u}^i(k)$, $i = 1, 2, \ldots, p$, by equation (8.16). Obtain $\hat{a}_i$, $\hat{b}_i$, $\hat{s}_{ij}$ from equation (8.17).

Step 4. Return to step 2 and repeat the calculations.

Step 5. After the above algorithm is converged, we can estimate γ_i from equation (8.12) by means of least squares since s_{ij} and b_j are now known quantities. The solution is

$$\hat{\gamma}_i = \left(\sum_{j=1}^{n} \hat{b}_j^2\right)^{-1} \sum_{j=1}^{n} \hat{b}_j \hat{s}_{ij} \quad i = 2, 3, \ldots, p.$$

The GLS estimates $\hat{a}_i$, $\hat{b}_i$, $\hat{s}_i$, and $\hat{c}_i$ are consistent. It follows that $\hat{\gamma}_i$ are also consistent. The GLS algorithm is readily applicable whenever off-line identification is called for. As for on-line identification, the GLS approach is tedious although it still can be implemented as described in chapter 6. It has been demonstrated that simpler on-line identification of Hammerstein model is possible when a learning model approach is used (Hsia and Bailey 1968).

8.6 Hammerstein Model—MSLS Identification

In this section, we present the multistage least-squares method for identifying the Hammerstein model. This method is a direct extension of the MSLS method described in chapter 7 for linear systems. The advantage of the MSLS method over the GLS method is that iterative computations are no longer necessary; also, the MSLS solution can be easily applied to on-line identification. However, MSLS parameter estimates are slightly less accurate even though they are consistent. Experimental results are presented later to show the actual quality of the MSLS estimates.

Rewrite equation (8.17) in the following form:

$$C(q^{-1})A(q^{-1})y(k) = C(q^{-1})B(q^{-1})u(k) + \sum_{i=2}^{p} \gamma_i C(q^{-1})B(q^{-1})u^i(k) + e(k)$$

Define

$$E(q^{-1}) = C(q^{-1})A(q^{-1}) = 1 + e_1 q^{-1} + \ldots + e_{m+n} q^{-(m+n)}$$

$$F(q^{-1}) = C(q^{-1})B(q^{-1}) = f_0 + f_1 q^{-1} + \ldots + f_{m+n} q^{-(m+n)}$$

$$R_i(q^{-1}) = \gamma_i F(q^{-1}) = r_{i0} + r_{i1} q^{-1} + \ldots + r_{i(m+n)} q^{-(m+n)}$$

Then

$$E(q^{-1})y(k) = F(q^{-1})u(k) + \sum_{i=z}^{p} R_i(q^{-1})u^i(k) + e(k) \qquad (8.18)$$

This is a difference equation having linear parameters e_i, f_i, and r_{ij}, and white residual $e(k)$. Thus we have a standard LS estimation problem. The MSLS solution is outlined below (Hisa 1968b).

Stage 1. Rewrite equation (8.18) as

$$y(k) = \boldsymbol{\phi}^T(k)\boldsymbol{\delta} + e(k)$$

where

$$\boldsymbol{\phi}^T(k) = [-y(k-1), \ldots, -y(k-m-n), u(k), \ldots,$$
$$u(k-m-n), u^2(k), \ldots, u^2(k-m-n), \ldots,$$
$$u^p(k), \ldots, u^p(k-m-n)]$$

$$\boldsymbol{\delta}^T = [e_1, \ldots, e_n, f_0, \ldots, f_{m+n}, r_{20}, \ldots, r_{2(m+n)}, \ldots, r_{p0} \ldots r_{p(m+n)}]$$

Compute the LS estimate $\hat{\boldsymbol{\delta}}$ by

$$\hat{\boldsymbol{\delta}} = \left[\sum_{k=1}^{N} \boldsymbol{\phi}(k)\boldsymbol{\phi}^T(k)\right]^{-1} \sum_{k=1}^{N} \boldsymbol{\phi}(k)y(k)$$

N is the data length.

Stage 2. From the exact relationship $r_{ij} = \gamma_i f_j$, $i = 2, 3, \ldots, p$, $j = 0, 1, \ldots, (m+n)$, we can estimate γ_i by least squares when the estimates $\hat{f}$ and $\hat{r}$ are substituted into the relationships. This yields

$$\hat{\gamma}_i = \left[\sum_{j=0}^{m+n} \hat{f}_j^2\right]^{-1} \sum_{j=0}^{m+n} \hat{f}_j \hat{r}_{ij} \quad i = 2, 3, \ldots, p.$$

Hence the nonlinear gain parameters γ_i are estimated.

Stage 3. To estimate a_i and b_i parameters, we can use the relationship

$$E(q^{-1})B(q^{-1}) = F(q^{-1})A(q^{-1})$$

and the estimates $\hat{E}(q^{-1})$, $\hat{F}(q^{-1})$ obtained from $\hat{\boldsymbol{\delta}}$. That is, we expand the polynomials on both sides of

$$\hat{E}(q^{-1})B(q^{-1}) = \hat{F}(q^{-1})A(q^{-1})$$

and then equate the coefficients of the like powers of q. This results in a set of $2n + m + 1$ linear equations of a_i and b_i with $\hat{e}_i$ and $\hat{f}_i$ as coefficients. Since there are only $2n + 1$ unknowns, we can readily estimate a_i and b_i by means of least squares.

Stage 4. This last step is employed to estimate the noise parameters c_i if they are called for. The basic relationships used for estimation are

$$\hat{E}(q^{-1}) = C(q^{-1})\hat{A}(q^{-1})$$
$$\hat{F}(q^{-1}) = C(q^{-1})\hat{B}(q^{-1})$$

Again a set of $2(m + n)$ linear equations of c_i can be obtained from the above when coefficients of like powers of q in those equations are equated. A set of least-squares estimates $\hat{c}_i$ can be readily obtained.

We have presented a four-stage least-squares estimation procedure to estimate two sets of system parameters and one set of noise parameters. Because the estimate $\hat{\delta}$ in stage 1 is consistent, it is easy to see that the estimates in the subsequent stages are also consistent. As pointed out in chapter 7, these MSLS solutions can be implemented in an on-line mode.

Two examples are now presented to demonstrate the performance of the MSLS identification method.

Example 1

In this example, we have chosen the following system structure:

$$x(k) = u(k) + 0.01625u^2(k) - 0.13295u^3(k) - 0.0061u^4(k)$$
$$A(z^{-1}) = 1 + 0.9z^{-1} + 0.15z^{-2} + 0.02z^{-3}$$
$$B(z^{-1}) = 0.42z^{-1} - 0.9z^{-2}$$
$$C(z^{-1}) = 1 + z^{-1} + 0.41z^{-2}$$

The system is excited by a Gaussian random sequence $\{u(k)\}$ with zero mean and unity variance. The noise source $e(k)$ is also a Gaussian random sequence of zero mean and variance $\sigma^2 = 0.09$. A data length of $N = 450$ was used for identification.

The MSLS identification results are listed in table 8–1. The sample mean and standard deviation are calculated from 10 independent computer runs. It is noted that in stage 1, the parameter vector δ has a dimension of 21, in which $f_0 = f_5 = r_{20} = r_{25} = r_{30} = r_{35} = r_{40} = r_{45} = 0$ because of the a priori knowledge of $b_0 = b_3 = 0$ in the polynomial $B(q^{-1})$. We see that the resulting system parameter estimates agree very well with their exact values. The parameters c_i are not computed.

Example 2

For this example, we keep everything the same as in example 1 except that the nonlinear gain element is changed to a saturation function given by

$$x(k) = \begin{cases} 1 & u(k) > 1 \\ u(k) & |u(k)| \leq 1 \\ -1 & u(k) < -1 \end{cases}$$

The nonlinear function is then approximated by the fourth-order polynomial

$$x(k) = u(k) + \gamma_2 u^2(k) + \gamma_3 u^3(k) + \gamma_4 u^4(k)$$

The corresponding identification results are listed in table 8–2. Since the parameters γ_i have no exact values to compare with, a plot of the estimated nonlinear gain is made in figure 8–4 for comparison with the ideal saturation function. The approximation is valid within the range of $|u(k)| \geq 3.0$. We note that the γ_i estimates and the range of approximation are dependent on the strength of the input $u(k)$. For example, if a larger range approximation of the nonlinear gain is desired, we need to use a random input of larger variance as well as a higher order polynomial of $x(k)$.

8.7 Concluding Remarks

In this chapter, we have discussed the problem of identifying nonlinear dynamic systems. One major problem in dealing with nonlinear systems is the lack of a general mathematical theory to describe various types of nonlinearities. Thus a specific system structure must be selected before parameter identification can be carried out. The Volterra series model and the Hammerstein models have been discussed in detail.

We have pointed out the difference in identification between the linear-in-parameter case and the nonlinear-in-parameter case. In the former, the linear LS techniques can be easily extended to identify the parameters. As we have pointed out, it is very difficult to estimate parameters in a noisy nonlinear system that involves nonlinearity in the output variable (Hsia and Ghandi 1972). The LS techniques are effective only for the class of nonlinear systems in which the output variables are linear, and in which the random disturbances are present only at the output.

Table 8–1
Parameter Estimates of Example 1 Based on 450 Points Input-output Data and 10 Computer Runs

Parameters	γ_2	γ_3	γ_4	a_1	a_2	a_3	b_1	b_2
True values	0.01625	−0.13295	−0.00616	0.9000	0.1500	0.0200	0.4200	−0.9000
Estimates	0.01047	−0.1337	−0.00526	0.9078	0.1540	0.01638	0.4277	−0.9088
Standard deviations	0.475×10^{-2}	0.240×10^{-3}	0.193×10^{-4}	0.233×10^{-2}	0.792×10^{-2}	0.310×10^{-3}	0.116×10^{-2}	0.219×10^{-2}

Table 8–2
Parameter Estimates of Example 2 Based on 450 Input-output Data Points and 10 Computer Runs

Parameters	γ_2	γ_3	γ_4	a_1	a_2	a_3	b_1	b_2
True values	—	—	—	0.9000	0.1500	0.0200	0.4200	−0.9000
Estimates	0.01644	−0.08686	−0.004788	0.9172	0.1734	0.01691	0.4076	−0.8477
Standard deviations	0.514×10^{-2}	0.142×10^{-3}	0.133×10^{-4}	0.282×10^{-2}	0.768×10^{-2}	0.185×10^{-2}	0.136×10^{-2}	0.222×10^{-2}

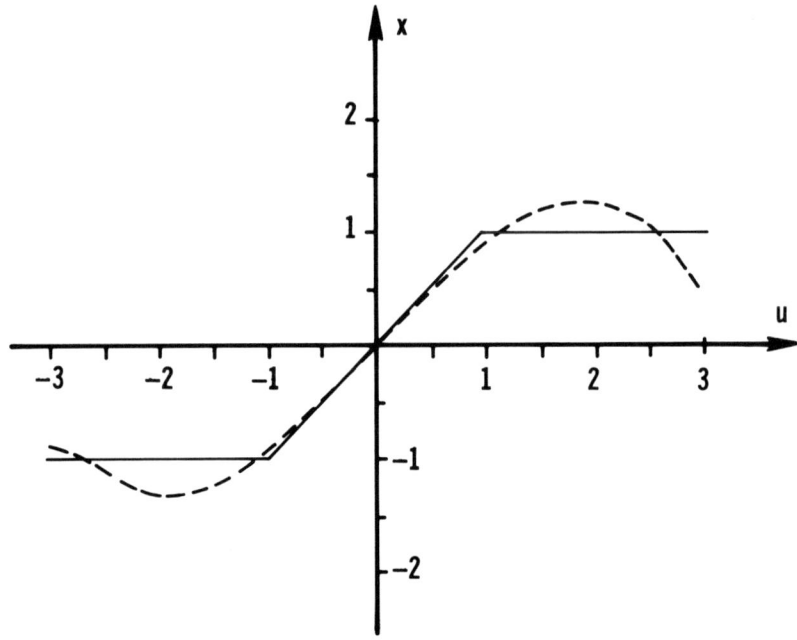

Figure 8-4. Comparison of the Exact and Polynomial-approximated Saturation Nonlinear Gain

References

Bányász, C.; Haber, R.; and Keviczky, L., "Some Estimation Methods for Nonlinear Discrete Time Identification," *3d IFAC Symposium on Identification and System Parameter Estimation,* pp. 93-802, 1973.

Chang, F. H. I., and Luus, R., "A Noniterative Method for Identification Using Hammerstein Model," *IEEE Transactions on Automatic Control*, Vol. AC-16, pp. 464-468, Oct. 1971.

Eykhoff, P., *System Identification*, Wiley, New York, 1974.

Haist, N. D.; Chang, F. H. I.; and Luus, R., "Nonlinear Identification in the Presence of Correlated Noise Using a Hammerstein Model," *IEEE Transactions on Automatic Control*, Vol. AC-18, pp. 552-555, Oct. 1973.

Hsia, T. C., "Least Square Method for Nonlinear Discrete System Identification," *Conference Record*, 2d Asilomar Conference on Circuits and Systems, pp. 423-426, 1968.

Hsia, T. C., and Bailey A. L., "Learning Model Approach for Nonlinear System Identification," *IEEE System Science and Cybernetics Conference Record*, pp. 228-232, San Francisco, 1968.

Hsia, T. C., and Ghandi, M., "An Iterative Algorithm for Non-linear Discrete System Identification," *International Journal of Control*, Vol. 16, pp. 763–775, 1972.

Hsia, T. C., "On a Least Squares Estimation Algorithm and Its Application to Digital Frequency Demodulation," *Proceedings ISA 22d International Instrumentation Symposium*, pp. 569–572, San Diego, 1976a.

⎯⎯⎯, "A Multi-stage Least Squares Method for Identifying Hammerstein Model Nonlinear Systems," *Proceedings 1976 IEEE Conference on Decision Control*, pp. 934–938, Dec. 1976b.

Inbar, G. F.; Hsia, T. C.; and Baskin, R. J., "Parameter Identification Analysis of Muscle Dynamics," *Mathematical Biosciences*, Vol. 7, pp. 61–79, 1970.

Narenda, K. S., and Gallman, P. G., "An Iterative Method for the Identification of Nonlinear Systems Using a Hammerstein Model," *IEEE Transactions on Automatic Control*, Vol. AC-11, pp. 546–550, July 1966.

Roy, R. J., and Sherman, J., "A Learning Technique for Volterra Series Representation," *IEEE Transactions on Automatic Control*, Vol. AC-12, pp. 761–764, Dec. 1967.

Schetzen, M., "A Theory of Non-linear System Identification," *International Journal of Control*, Vol. 20, pp. 577–592, Oct. 1974.

Volterra, V., *Theory of Functionals and of Integral and Integro-Differential Equations*, Dover, New York, 1959.

Index

Index

Additive output noise, *127*, 131
Adjustment gain, 57-58
Algorithms, 4, 29-31, 81, 83, 93
   for complete system identification, 81, *82*
   for generalized least squares, 101-106, *104, 105*, 111-116
   gradient, 57
   for Hammerstein model, 148-149
   Kalman filtering, 84, 93-95
   for on-line least squares, 55-58, *58*, 81
   recursive (*see* Recursive algorithms)
   sequential, 22-25, 29-31, 81, 83-84
   updating, 23
Anderson, G.W., 37, 43, 45
Åström, K.J., 122
Autocorrelated random process, 98
Autocorrelation function, 43-45
   of pseudo-random binary sequence (PRBS), 49-51, *51*

Bányász, C., 122, 146
Basic system identification problem, 68
Batch processing estimation results, 55
Best linear unbiased estimator, 21
Bias
   in least-squares estimation, 20, 72-73, 98-100
   with linear-parameter nonlinear difference equation, 143
   of multistage least-squares estimator, 127
   of pseudo-random binary sequence autocorrelation function, 49
   of weighting sequence, 41
Black box problem, 2
Bohlin, T., 122
Buland, R.N., 37, 43, 45

Canonical forms, 11-14
Chang, F.H.I., 148
Clarke, D.W., 81, 111

Coefficients, regression, 18
Complete identification problem, 2, 3, 68
Computer techniques, *54*, 67
   advantages of using PRBS test signal, 51
   difficulties, minimization of, 43, 60-61, 111, 120, 125, 129
   program for PRBS generation, 65
   time use, 136-138
Consistency
   of generalized least squares, 108-109
   of instrumental variable estimator, 120
   of least squares, 21, 73
   of multistage least squares, 127, 132, 134
   of nonlinear system estimator, 144, 149
Continuous representation of weighting function, 62
Continuous systems, approximation of, 37, 67, 84, 86
   identification, 84, 86-89, *88*
Continuous-time models, 7
Control systems, design of, 1
Convergence rate, improvement of, 118
Convolution integral, 38
Convolution summation, 9-10, 11, 39
Cooper, G.R., 37, 43, 45
Correlated residuals, 98-100
Correlation functions, 43-45
Cross-correlation functions, 43, 45, *46*, 51

Davies, W.D.T., 49
Derivative signals, substitutes for, 88-91
Difference equations, 7, 8
   compared to differential equations, 86-89

Difference equations *(continued)*
  "enlarged," 132-133
  relation to canonical form, 13-15
Differential equation models, 84, 86.
  See also Difference equations
Direct identification approach, 67
Discrete-time models, 7, 67
Discrete-time weighting sequence identification, *40*
Dynamic system models, 7

Efficiency
  of generalized least squares, 108-109
  of instrumental variable estimator, 121
  of least squares, 21
Equation error, 4, 68, *69*. See also Generalized equation error
Error, in estimation techniques, 4
Error covariance, 20-21, 41-42, 47, 73
  measure of, 24
  minimization of, 21
Error function
  exponentially weighted, 81, 83
  influence of noise level on, 106
Error signal, *69*
Estimation
  in basic system identification, 68-72
  sequential, 22, 24-25, 29-31, 81
  of weighting functions, 37, 39
Eykhoff, P., 58, 141

Filtered white noise, 100
Filters, state variable, 88-89
Filter transfer function, 100
Fitting error, 68, 89-90
  noise interpretation of, 73-74
  weighting of, 24

Gallman, P.G., 148
Generalized equation error, 101, *102*, 109
Generalized least-squares estimation
  algorithm, 101-106, *104, 105,* 111-116, 118-119
  applied to Hammerstein model, 148-149
  convergence rate of, 106, *107,* 118
  on-line version for, 106-108
Generalized least-squares method, 97, 100-101, 108-109
  compared to multistage least squares, 125, 136-138, 149
Gertler, J., 122
Gobring, B., 76
Goodness-of-fit comparison method, 76-77
Goodwin, G.C., 47
Gradient algorithm, 57
Graupe, D., 37, 58
Gray box problem, 2

Haber, R., 146
Haist, N.D., 148
Hammerstein model, 146-152, *147*
Hasting-James, R., 106
Hill, J.D., 53, 54
Hsia, T.C., 81, 84, 89, 111, 113, 132, 135, 144, 148

Identifiability, 42-43, 51, 73
Identification. *See* System identification
Initial values, selection of, 25
Input
  in linear dynamic systems, 38, 42
  as sequence of stationary variables, 45
  signal, 47-48
  as white random process, 45
Instrumental variable method, 119-122, *121*
  on-line version, 121-122
Iserman, R., 130

Kalman filtering approach, 84, 93-95
Kalman, R.E., 93
Keviczsky, L., 146

Learning model identification, 57, *58*
Least-squares estimation, 19, 55
  conditions for existence of, 72

of noisy system model, 98-99, 148
as number of parameters increases, 27-29
of parameter vector, 70
recursive algorithm for, 22-25
of single-variable linear time-invariant discrete system, 69-72
statistical properties of, 20-22, 72-74
weighted, 19
Least-squares identification method, 4
digital orientation of, 36, 37
using pseudo-random binary sequence, 69-72
Least-squares solution, 43-46, *46*
Least-squares theory, 17-19
generalization of, 25-27
Levin, M.J., 37, 41, 48, 62
Likelihood equation, 33-34
Linear discrete dynamic systems, 9, 67-80
Linear-parameter nonlinear difference equation, 143
Lion, P.M., 89
Luus, R., 148

McMurty, G.J., 53, 54
Markov estimator, 21, 108
Matrix inversion lemma, 23-24, 35
Maximum-length sequence, 49
Maximum likelihood estimator, 33-34
identity to least-squares estimator, 22, 34, 73
of weighting sequence, 42
Measurement noise. *See* Noise
Mehra, R.K., 47
Mendel, J.M., 37, 57
Minimal parameter representation, 11-14
Minimum-error-squares method, 18-19
Minimum-output-error identification, 109-110, *110*
Minimum variance estimator, 21, 41
Model errors, independence of, 77
Model follower techniques, 89
Model order determination. *See* Order determination

Multi-dependent-variable systems, 25-26, *26*
Multistage least-squares method, 125-135, *126*
applied to Hammerstein model, 149-152
compared to generalized least squares, 125, 136-138, 149
employing weighting sequence, 126-127
identification procedures, 126-127
Multivariable linear continuous system identification, 58-61, *60*
Multivariable linear discrete system, *10, 12*
Murdoch, J.C., 47

Nagumo, J., 37, 57
Narendra, K.S., 148
Noda, A., 37, 57
Noise, 38
disturbance, 31, 89-90, 148
disturbed system equation, 19
effects, 2, 38, *39*
in estimating weighting functions, 37
Noise transfer function
in generalized least squares, 105, 110-111
in multistage least squares, *127*, 129-130, 131
Noisy system model, 97-101, *98, 102*
Nonlinear systems, 146, *147*
identification problem of, 141-152, *142*
minimization problem of, 111
representation, 143
single input-single output, 141
Nonparametric models, 37, 67, 141-143
Normal equation, 19

Off-line identification, 4, 55, 149
One-shot estimation results, 55
On-line estimation. *See* Sequential estimation
On-line identification, 4-5, 55-58, *58*,

On-line identification *(continued)*
  75, 81, 149
  application of, 81, 83-84
  of generalized least squares, 106-108, 116-119
  of instrumental variable solution, 121-122
  of system-noise parameters, 116-119, *117*
On-line steepest decent techniques, 89
Optimum input signal, 47-48, 72
Order determination, 76-81
Ordinary least squares, 19
Output, in linear dynamic system, 38
Output additive disturbance, 109-110
  error, 4, 57, 109
  measurement noise, 73-74, *74*

Pagurek, B., 126, 129
Pandya, R.N., 122, 126, 129
Parameter estimation, 4, 18-19, 68-72
  with continuous system, 87-88, 89
  methods compared, 136, 138
  in multistage least squares, 130-132
  reduction in, 118
Parameter identification, 2, 4, 84, 94
  nonlinear, 144-145, 146
Parameter number increase, 27-29
Parametric models, 37, 67, 143
Payne, R.L., 47
Period choice, 49, 51
Persistently exciting inputs, 42, 70
Polak, E., 119, 122
Pseudo-random binary sequence (PRBS), 49-55, *50, 51*, 72, 129
  generation of, 49, *52*, 65

Random process, white. *See* White random process
Real-time identification algorithm, 29-31, 81, *85, 86*, 93
Recursive algorithms
  for generalized estimation, 118-119
  for least squares, 22, 24-25, 27-29, 76-81
  for multistage least squares, 129, 135

parameter updating, 81, 83
for weighting function identification, 55-58
Recursive parameter estimation, 93
Recursive solutions, 22
Regression function, 18
Representation problem, 4
Residuals
  autocorrelated, 97
  in linear-regression theory, 19, 68
  sequence, 77
  uncorrelated, 68, 98, 132
Roy, R.J., 142

Sage, M.W., 106
Schetzen, M., 141
Settling time parameter, 43, 49, 51
Sherman, J., 142
Single-variable systems, 7-8, *8*, 68, *69*
Smith, F.W., 84
Söderström, T., 106, 111, 122
State variable equation, 11-15
State variable filters, 88-89
Stationary independent variables, 45
System equation, 87
System identification, 1, 3, 37
  algorithm, 81, 82
  problem, 1-3, 68
  results, 56
  techniques, 7, 40-45, 49-61, 126-135
  of weighting function, *39*
System-noise identification problem, 125, 144
System-noise model, 97-101, *98, 102*
System representation, specific, 141
System structure determination. *See* Order determination

Talmon, J.K., 116
Test signal, white noise, 49-52, *50, 51*, 54, 72
Time-lag estimation, 81
Time-varying parameters, 29, 81, 83, 84
Time-varying systems, 5, 56
Transfer function, 8-9, 68

defining Z-transform, 8
filter, 100
in generalized least squares, 101, 105, 110-111
and noise. *See* Noise transfer function
Turin, G.L., 37

Unbehanen, H., 76
Uncorrelated residual, 68, 98, 132

Van den Boom, A.J.W., 116, 134
Van Den Enden, A.W.M., 134
Volterra, V., 141
Volterra series, 141-142

Weighting, equal, 29
Weighting function, 37, 38, 62
identification, 40-43, 54-58, 61
matrix, 58-59, 60-61
representation of Volterra Kernel, 141

samples, 43-45
Weighting matrix, 10-11
Weighting schemes, 29-30
Weighting sequence, 9-10, 39
estimator, 41-43
matrix, 15
model, 67
in multistage least squares, 126-127, 130-132
Z-transform of, 10
Whitening filter, 100-101
White random process
fitting error as, 90
output additive disturbance as, 109-110, *110*
system input as, 45, 47-48
White residual, 68, 98, 100
Wiener, N., 45
Wiener-Hopf equation, 45
Wieslander, J., 93, 95
Wittenmark, B., 93, 95
Wong, K.Y., 119, 122

Young, P.C., 122

About the Author

T.C. Hsia received the BEE from National Taiwan University, and the Ph.D. from Purdue University. He has been on the faculty of the Department of Electrical Engineering at the University of California, Davis, since 1965, where he is currently a professor.

Dr. Hsia's teaching and research activities are in the fields of control theory, system identification, pattern recognition, and biomedical engineering, and he has published over fifty research papers on these topics. He is also an industrial consultant in digital signal processing and pattern recognition.